**50 experiment
with renewable en**

The Franzis Tutorial Kit

50 experiments with renewable energy

FRANZIS

Dear customers!

This product was developed in compliance with the applicable European directives and therefore carries the CE mark. Its authorized use is described in the instructions enclosed with it. In the event of non-conforming use or modification of the product, you will be solely responsible for complying with the applicable regulations. You should therefore take care to assemble the circuits as described in the instructions. The product may only be passed on along with the instruction and this note.

© 2014 Franzis Verlag GmbH, Richard-Reitzner-Allee 2, 85540 Haar, Germany

Author: Ulrich E. Stempel
Translation and DTP: G&U Language & Publishing Services GmbH
Art & Design Cover: www.ideehoch2.de

ISBN: 978-3-645-65233-9

Contents of the Tutorial Kit

Before you begin with the experiments, check the components for completeness according to the following list.

Components included in the tutorial kit:

Number	Component	Specification
1	solar module	crystalline
1	breadboard	SYB-46
1	gold cap	1 F, 2.7 V
1	motor/generator	3 V
1	cardboard disc, printed	
1	moving-coil gauge	scale for displaying V, mA, °C, m/s
1	capacitor	10 nF
1	electrolytic capacitor	4700 µF, 6.3 V
1	LED	5 mm, red
1	LED	5 mm, orange
1	LED	5 mm, red, flashing
1	wire	diameter 0.6 mm, length 0.8 m
10	resistors (carbon)	various resistance values
1	trim potentiometer	10 kΩ
1	silicon diode	1N4148
1	transistor	2N3904

Number	Component	Specification
1	IC	LM 358
2	NTC	4.7 kΩ
2	cables with alligator clips	red/black
4	pins	
1	battery clip	for a 9-V battery
1	push-button	
1	brass adapter ring (for the motor shaft)	
2	cut-out cardboard templates, printed	
2	plastic bags, transparent	
2	single-use syringes	10 ml

In addition, you will need the following common household items:
- plastic water bottle 0.5 l
- wrapping film
- sun glasses (eye protection)
- scissors
- glue, double-sided adhesive tape
- 9-V battery

Table of Contents

Preface

Energy provided by nature has played a major role in the history of mankind and it is becoming more and more important to supply the world with renewable energy. The political aspect of using locally developable renewable energy is of increasing significance. In particular non-industrialised nations can gain independence and autonomy by utilising renewable energy and thus detach themselves from the existing power structures. This could end the global conflicts about energy and raw materials.

Utilising the power of sunlight, wind and water has always been a concern of mankind. Over the ages people have used wind and water mills for grinding grain and they also had saw mills driven by water power. Even today many old water power stations are still in use, mainly to produce electricity.

Experimenting with renewable energy may seem easy at first sight. However, when it comes to putting it into practice, it is important to gain some basic knowledge beforehand. The experiences and the knowledge you gain with the models presented in the experiments can be applied to and used in larger real-life constructions.

Ulrich Stempel, January 2011

1 The Components

1.1 Breadboard

The breadboard features internal contact springs that are connected in series. It is highly suitable for implementing electronic circuits for renewable energy systems. The electronic components und jumper wires can be inserted into the contacts as often as you want. This way you can set up the circuits without soldering or screwing and experiment with the set-up just by rearranging or replacing individual components.

The breadboard included in the tutorial kit has a total of 270 contacts in a 1-inch grid. The 230 contacts in the middle section are connected to rows of five by vertical strips.

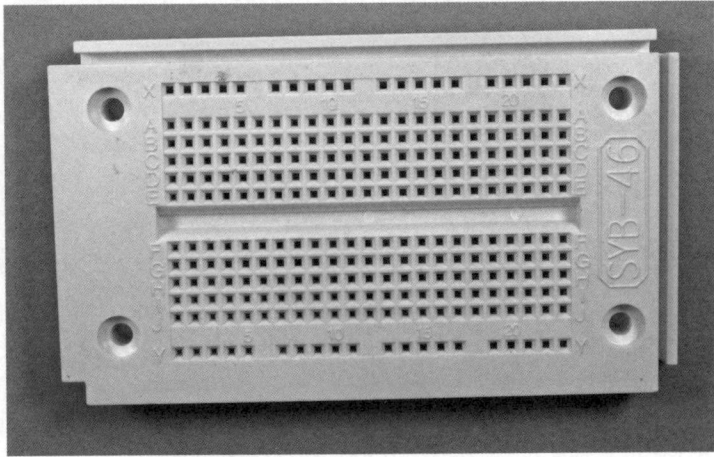

Figure 1.1: Breadboard

On the edge of the long side there is a row with 20 contact points that are horizontally interconnected via a bar. These 'upper' and 'lower' rows are suited for power supply.

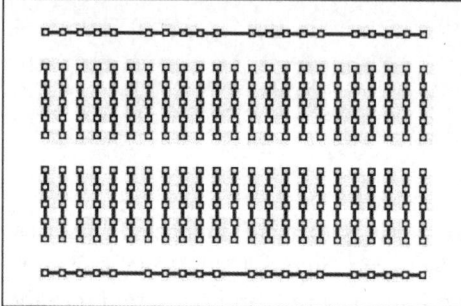

Figure 1.2: Layout of a breadboard

Inserting jumper wires and the connecting leads of components into a new breadboard can be tedious. You can mitigate this problem by inserting a thin fixing pin into the contacts. The contact springs are intended for wires with a diameter of 0.3 to 0.6 mm. The needle should not be too thick because otherwise, the contact springs will wear out too much. This in return would impair the contact to the inserted leads.

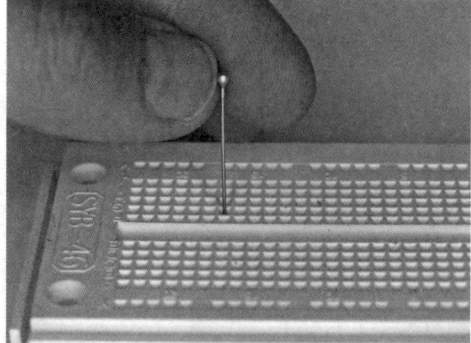

Figure 1.3: You can widen the contacts with a pin.

The following building instructions recommend to connect the lower bar to the negative pole of the power supply and the upper bar to the positive pole. All components of the tutorial kit can be directly inserted without soldering. Since the connecting leads tend to buckle, you need some practice inserting the components. Grip the connecting lead at a short length and insert it vertically and with little force into the contact point.

The connections of the solar module and the motor can be stabilised with additional pins.

Figure 1.4: Connecting jumper wires from solar module, motor/ generator etc. with additional pins

Pinching off the ends of jumper wires with a wire cutter in an oblique fashion facilitates the insertion. Very short leads like those of a transistor can be inserted with small flat nose pliers or forceps so that they do not bend. Jumpers can be made out of the included wire with 0.6 mm diameter. Estimate or measure the required length (plus some length for the ends that have to be inserted into the contacts). Remove the insulation at the ends, using a precise cable stripper or by cutting into the insulation all around the wire and stripping it off.

1.2 The solar module

The included solar module consists of several crystalline solar cells that are mounted on a baseplate and coated with a transparent protective layer. The silicon substance of the individual solar cells consists of several crystals. A solar cell works as follows: Pure silicon is purposefully contaminated so that a negative and a positive layer are created. On top is the (negative) N layer. In order to enhance the absorption of light, an additional dark blue coating is applied to the surface. The lower layer is the P layer. When light is shed on the cell, the electrons begin to move and thereby a voltage occurs between the two layers.

We can use this voltage and the flowing current. A single crystalline silicon solar cell generates a voltage of up to 0.5 V. The current depends on the size of the cell.

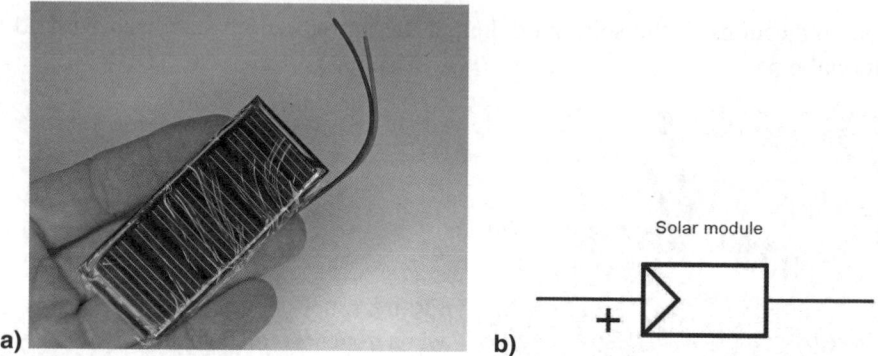

Solar module

a)   b)

Figure 1.5: (a) Solar module (b) Symbol in circuit diagrams

1.3   Electrical machine (motor/generator)

The tutorial kit contains a permantly excited DC motor that you can use in your experiments as a motor or as a generator. Technically speaking, there is no difference between a motor and a generator. Whether an electrical machine works as a motor or as a generator depends on the given application and wiring. You can therefore use the included electrical machine to drive something as well as to generate power.

The permanent excitation of a generator has the advantage that the required field for induction/current generation is not created by a coil but by a permanent magnet. Therefore no energy or electric current is needed for the field. Especially in the case of wind- and water-driven low-speed generators, this field would consume more energy than what it gains. The drawback is that it is harder to start up than a separately excited generator, due to the 'gluing' effect of the magnets and the friction of the carbon brushes.

When you use this machine as a motor, however, it already starts up at very low currents and low voltages.

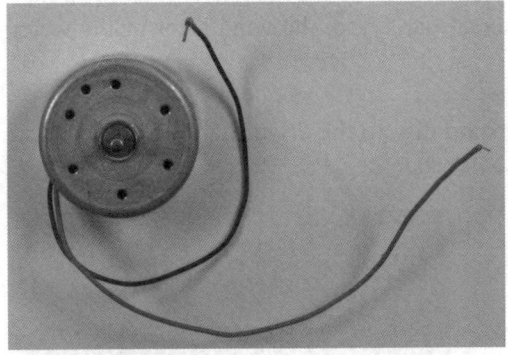

Figure 1.6: DC machine

M = Motor

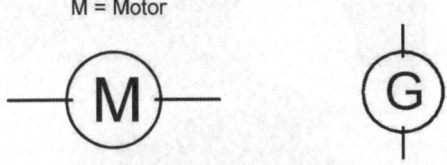

Figure 1.7: Symbol
for a generator/motor

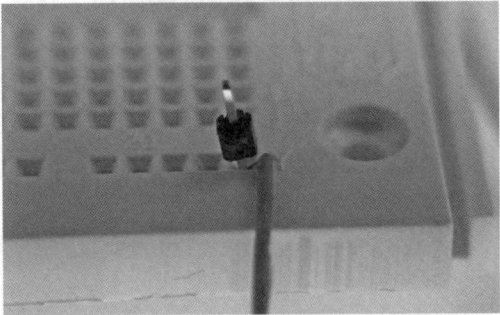

Figure 1.8: The connection
leads of the generator/motor
consist of flexible wire. The kit
contains some pins in order to
simplify the connection with the
breadboard.

The connection leads are made of flexible wire. By using the included pins, you can connect them without any difficulty with the breadboard. To this purpose, you first insert the bare wire ends into the contact springs and then secure them with the pins. A hard tool like e. g. the flat blade of a screwdriver may be helpful.

1.4 Adapter ring and cardboard disc

The tutorial kit contains an adapter ring in order to mechanically join the motor/generator drive shaft with supplemental parts, e. g. the wind rotor or the water wheel.

When you attach the included cardboard disc to the drive shaft of the electrical machine, you can see whether the shaft rotates.

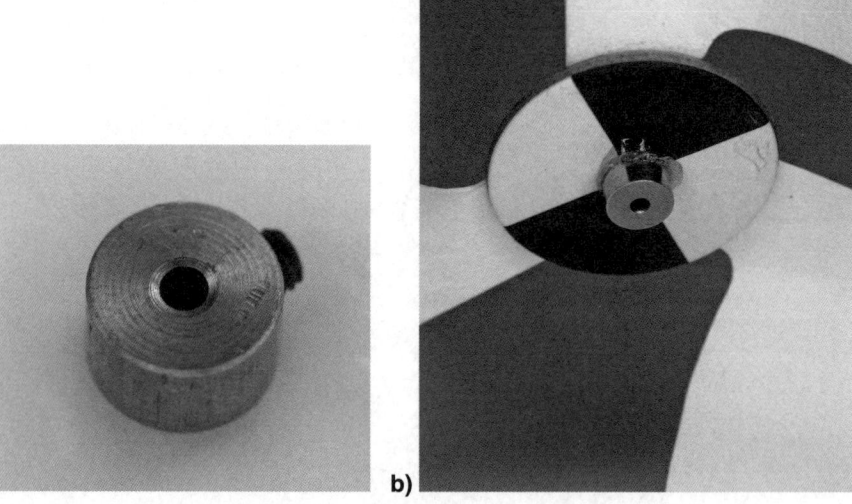

a)   b)

Figure 1.9: (a) Adapter ring (b) Glued to the cardboard disc

1.5 Moving-coil gauge

A moving-coil gauge is the classical measuring device for electrical current, voltage and related values. It consists of a permanent magnet and a rotating iron cylinder between its poles that is wrapped in a coil. When an electrical current flows through the coil, the needles of the gauge are deflected. The extent of this deflection depends on the intensity of the current. When you connect the gauge according to the imprinted polarity, the needle is deflected into the right direction. When you connect it the other way round, the needle will hit the left stop.

When you calibrate the gauge and label the scale, you can directly read off the current value. The coils are only suited for very small currents. It is possible to measure higher currents by using appropriate additional resistors (shunts) through which the main part of the electrical current flows. The moving-coil gauge measures primarily the current and can therefore be used as an ampereme-ter. To this end it is connected in series in the circuit.

In the experiments of this tutorial kit, the moving-coil instrument is also used for other measuring tasks. Therefore you will find labels for the measuring ranges like V (voltage in Volt), mA (current in milliampere), C° (temperature in °C) and m/s (wind speed in metres per second) below the scale. The external wiring of the instruments determines the respective display.

Figure 1.10: Moving-coil gauge with scale

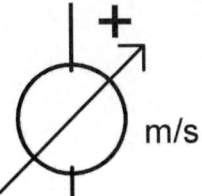

Figure 1.11: Symbol with additional indication of the measuring range, here metres per second

Note

The measuring instrument must not be directly connected to a battery because it could be damaged.

1.6 Diodes

Diodes let pass electrical current only in one direction. Among other things, they are used to rectify alternating currents and to block unwanted polarity in direct current.

The operation of a rectifier diode like those that are included in the tutorial kit correlates to that of a check valve. When you apply pressure (voltage) to the valve (diode) in the reverse direction, the electrical current is blocked. In the other direction (as indicated by the arrow), the pressure (voltage) must be high enough to overcome the spring pressure (block voltage). After that, the valve (diode) opens and the fluid or gas (current) can flow. The required pressure to overcome the spring pressure in the mechanical model correlates to the so called *forward bias* of the diode. A specific voltage must be applied to the diode in flow direction so that the diode becomes conductive.

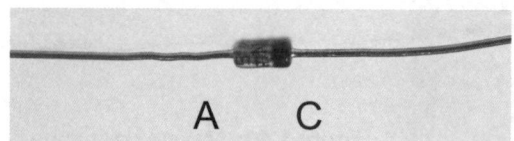

Figure 1.12: A silicon diode of type 1N4148. The cathode is descernible by the imprinted black ring. The other connection lead is the anode. The technical direction of electric current runs from the anode to the cathode.

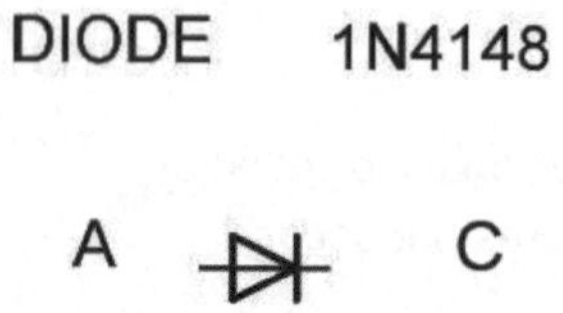

DIODE   1N4148

A ⫤▷⊢ C

Figure 1.13: Symbol for a diode

In silicon diodes (e.g. 1N4148), a noteworthy current in flow direction (as indicated by the arrow in the symbol) appears not until a voltage of approx. 0.6 V to 0.7 V is reached (700 mA). In Schottky diodes, the current already begins to flow at approx. 0.25 V.

1.7 LEDs

In addition to the properties of standard diodes, light emitting diodes (LEDs) exhibit another feature: They light up when a voltage is applied. The tutorial kit contains a red, an orange and a red flashing LED. You can tell the flashing LED by the small black dot inside the red casing. This is the integrated circuit that prompts the LED to flash.

Always use LEDs with a series resistor. You can calculate the required resistance value of this component by the formula R = V/I (R = resistance in Ohm, V = voltage in Volt, I = current in Ampere).

Example: A normal LED (red, orange, green, white) needs approx. 20 mA to shine brightly. When you have a voltage of 9 V, you have to divide it by 0.02 A (20 mA), yielding 450 Ω. The tutorial kit includes resistors in the range of 10 Ω to 10 kΩ. In our example, you can use two 1-kΩ resistors. When you connect them in parallel, you get a resistance of approx. 470 Ω.

For most of the experiments with the solar module, a series resistor of 100 Ω is suitable. With higher resistance values (e.g. 1 kΩ) the LED gets a significantly lower current and does not shine as brightly.

Contrary to a light bulb, an LED has no filament. Therefore is has a longer service life and a lower current consumption. The reverse LED effect is used in solar cells. LEDs, too, can generate a small current when illuminated by a powerful light source. You can measure this effect only when you connect an electrolytic capacitor (completely discharged) parallel to the LED as a temporary buffer store. After some hours (depending on the light source), the LED has charged the capacitor. You can then measure the voltage with a multimeter.

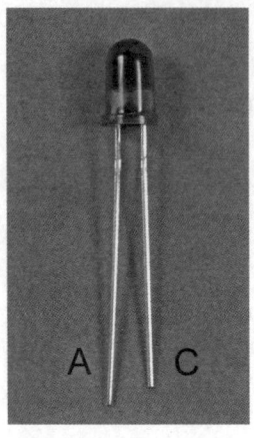

Figure 1.14: Connections of an LED: anode (+) with a longer connection lead, cathode (-) additionally marked by a bevel on the casing

LED, 5 mm

Figure 1.15: Symbol for an LED

Figure 1.16: Blinking LED with built-in blink IC

1.8   Transistors

Transistors are active components that are used to control and amplify current and voltage in electronic applications.

The bipolar transistor included in this tutorial kit is of the type 2N3904. It is a small-power transistor suitable for a maximum operating voltage of 30 V and a maximum current of 200 mA. The labels *N* and *P* identify the negative or positive semiconductor layers of the transistor.

E   B   C

Figure 1.17: Transistor connections: emitter (E), base (B) and collector (C)

A transistor works as follows: A low current applied to the base-emitter section can control a high current in the collector-emitter section. Thus, when you have a small base current (positive in NPN transistors, negative in PNP transistors), the transistor directs the current from the collector to the emitter or vice versa. When there is no current applied to the base or when the base is connected to a negative (NPN) or positive potential (PNP), the transistor blocks.

T1   2N3904

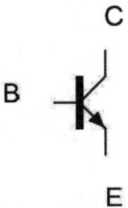

C

B

E

Figure 1.18: Symbol for an NPN transistor

1.9   Integrated circuits/op-amps

Integrated circuits (ICs) contain many electronic components on a minimum of space. One variant is the integrated circuit LM358 with two identical operation amplifiers (op-amps).

Fig. 1.19 shows the connections, Fig. 1.20 the inner layout of the IC. Both op-amps can be used independently. The supply voltage for both is applied to the same pins (pin 4 = minus, pin 8 = plus).

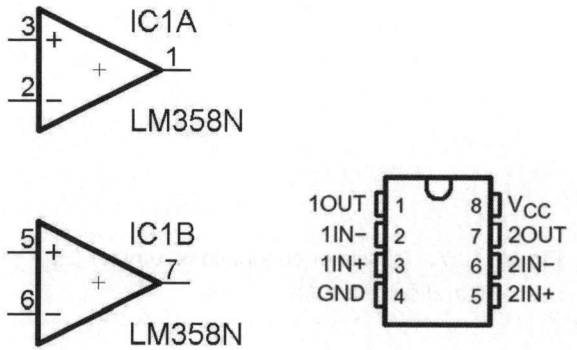

Figure 1.19: Symbol and pin layout of an LM358 IC

The pin layout of the IC can be determined by a notch at one end of the IC casing. The IC is usually inserted into the breadboard with the notch showing to the left. When you insert an IC for the first time, bend the legs a little between your thumb und your index finger so that they slip into the contacts more easily. To simplify the removal of the IC, you use a small screwdriver as a lever between the breadboard and the IC casing.

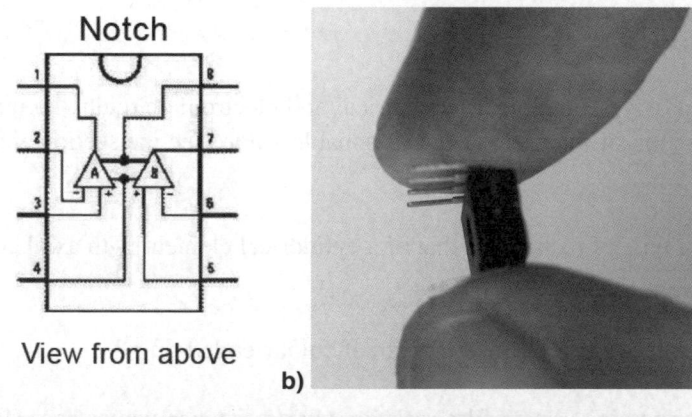

Notch

View from above

a)

b)

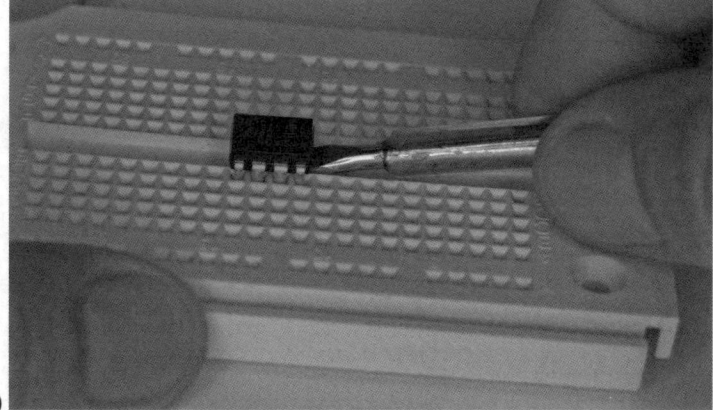

c)

Figure 1.20: (a) Internal wiring of the IC (b) Pinch the pins before inserting the IC (c) Removal of the IC

Note

Apart from the power supply, the IC is relatively insensitive to the external wiring. It is therefore recommendable to check the correct connection of the power supply. Reverse connecting the positive and negative poles usually leads to the destruction of the component. Always make sure that pin 4 is connected with the minus bar of the breadboard and pin 8 with the plus pole before you attach the circuit to the battery.

1.10   Resistors

A resistor is a passive component in electrical and electronic circuits. Its main task is the reduction of the current to a reasonable value (see the section about LEDs).

The best-known type of resistors is that of a cylindrical element with axial connection leads.

The resistance values are imprinted in form of colour-coded rings.

The tutorial kit contains carbon film resistors with resistance values according to the following table.

Quantity	Resistance value	Ring 1 Digit 1	Ring 2 Digit 2	Ring 3 Multiplier	Ring 4 Tolerance
1	10 Ω	brown	black	black	gold
1	100 Ω	brown	black	brown	gold
6	1 kΩ	brown	black	red	gold
1	2.2 kΩ	red	red	red	gold
1	10 kΩ	brown	black	orange	gold

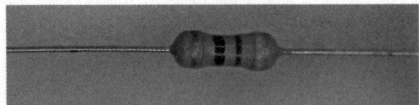

Figure 1.21: A resistor

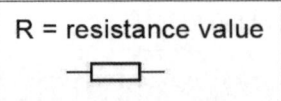

R = resistance value

Figure 1.22: Symbol for a resistor

Apart from the carbon film resistors included in the tutorial kit with a 5% tolerance you can as well use metal film resistors with a 1% tolerance for the experimental circuits presented in this kit. These types of resistors use a different system of colour-coding: ring 4 is used as the multiplier while ring 5 (not shown in the table) represents the tolerance.

Resistance value	Ring 1	Ring 2	Ring 3	Ring 4
10 Ω	brown	black	black	gold
100 Ω	brown	black	black	black
1 kΩ	brown	black	black	brown
2.2 kΩ	red	red	black	brown
10 kΩ	brown	black	black	red

1.11 Trim potentiometer (trimmer)

A potentiometer is a continuously variable resistor. The trim potentiometer in the learning kit has a maximum value of 10 kΩ and can be adjusted continuously between 0 and 10 kΩ with a screwdriver. Inside the component there is a resistor lead and a sliding contact or wiper to vary the resistance value. The resistor lead and the wiper are only suited for small currents. Consumers like LEDs and the motors must not be controlled directly by the trimmer because this would destroy the component.

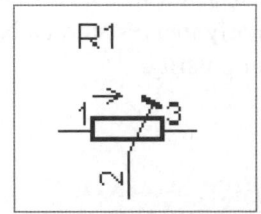

Figure 1.23: Trim potentiometer   **Figure 1.24:** Symbol for a trim potentiometer

Figure 1.25: Adjusting the trim potentiometer with a screwdriver

1.12 NTC

NTC (negative temperature coefficient thermistor) is the abbreviation for a high-temperature conductor i. e. a temperature sensor in which higher temperatures enhance the conductivity or reduce the resistance, respectively. The opposite of a high-temperature conductor is a *posistor* (PTC resistor) with a higher conductivity at lower temperatures.

The tutorial kit contains NTCs with a resistance value of 4.7 kΩ at 20 °C, aptly named NTC 4.7. Similarly to a resistor, an NTC can be connected both ways. It is a component without polarity.

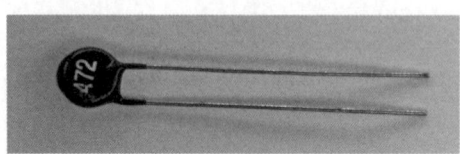

Figure 1.26: NTC sensor

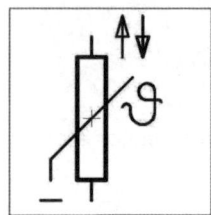

Figure 1.27: Symbol for an NTC

1.13   Capacitors

Capacitors consist of two metal plates and an insulation layer. When you apply a voltage, an electrical field builds up between the capacitor plates and stores energy. A capacitor with a large plate surface and a small distance between the plates can store a greater charge at a given voltage. The capacitance of a capacitor is measured in farad (F). There are different types of capacitors for various purposes and various capacitance ranges. Capacitors suited for high-frequency applications have a capacitance between 1 pF (picofarad) and 100 nF (nanofarad). They are built as ceramic capacitors in disc shape. You can see a capacitor of this type in the illustration.

The capacitance values are represented by three digits as follows:
First number = first digit of the value, second number = second digit, third number = multiplier

Examples:
Inscription 221 represents 220 pF
Inscription 102 represents 1 nF
Inscription 104 represents 100 nF

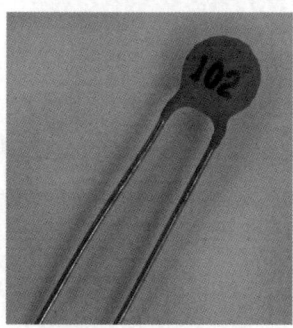

Figure 1.28: Ceramic disc capacitor

CAPACITOR

Figure 1.29: Symbol for a capacitor

1.14   Electrolytic capacitor

Electrolytic capacitors have a higher capacitance than ceramic capacitors. Imagine a capacitor in the original construction with two metal plates (as it is indicated in the circuit diagram symbol). The first 'plate' of an electrolytic capacitor is insulated by an oxide layer (dielectric medium), the second one consists of an electrolyte (a conducting fluid). Because of the electrolyte, an electrolytic capacitor has a polarity. Its connection leads are labelled as plus and minus pole. Using the component for a longer time period in the wrong direction will destroy the electrolyte.

Internally, an electrolytic capacitor is not composed of metal plates but of thin metal films that are wound up and incorporated into the casing, thus yielding a cylindrical shape with two connection leads. There are radial and axial variants.

Never exceed the indicated maximum voltage because this would destroy the insulating layer.

The tutorial kit includes a radial electrolytic capacitor with a capacitance of 4,700 µF suited for a maximum voltage of 6.3 V.

> **Note:**
>
> µF stands for 'microfarad'. The prefix µ denotes the millionth part of the base unit.

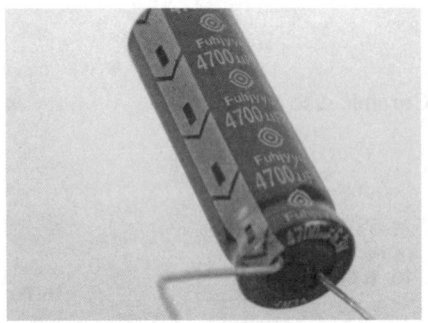

Figure 1.30: Electrolytic capacitor with connecting leads. The longer one is the plus pole. The minus pole is additionally marked by a light-coloured bar on the casing.

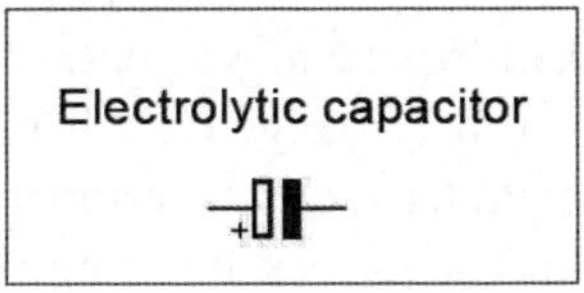

Electrolytic capacitor

—◻️▮|—

Figure 1.31: Symbol for
an electrolytic capacitor

1.15   Gold cap

Extremely high capacitances can be reached with gold caps. This name is a trademark of Panasonic. 'cap' is short for capacitor, but 'gold' is only a verbal gilding of the capacitor because there is no real gold incorporated in the component.

The gold cap stands between electrolytic capacitors and rechargeable accumulators. It made several storage technologies possible, especially in the field of solar technology, because, despite its high capacitance value it is very small. The maximum voltage is not particularly high and just accounts for some volts, but because of its high capacitance, the gold cap is suited as an interium power supply. It is especially useful in devices where data have to be stored while the device is switched off.

A gold cap has a maximum service life of 8–10 years that is even reduced when the operating temperature exceeds the approved value or when high electrical currents are often withdrawn. The capacitance is reduced gradually. A gold cap works best when it is discharged rarely and only with small current. A permanent wrong polarity will destroy the gold cap, but it will not explode.

The inner structure of a gold cap can be imagined as a parallel connection of many small electrolytic capacitors. When charging the gold cap, part of these capacitors are charged and subsequently charge the capacitors that reside further back. It is therefore important to wait long enough until the gold cap is completely charged (ten minutes to one hour). A gold cap that is only partially charged will immediately lose its charging voltage after the power supply is switched off!

Since a gold cap cannot be overcharged there is no need for a charging circuit with overcharge protection. You just charge it by a constant voltage until it is full. In this process it behaves like a capacitor. You do not have to worry about discharging. With this component, there is no such thing as a deep discharge. Completely emptying a gold cap does not impair its service life.

Between the many individual capacitors reside series resistors that yield a high inner resistance. The application of this type of capacitor is therefore limited. It is not suited for smoothing pulsating DC voltages or as a coupling capacitor.

The minus pole is indicated by a white bar and has a shorter connection lead.

Figure 1.32: Gold cap

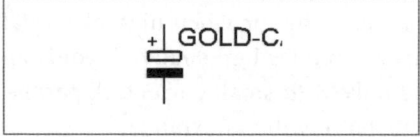

Figure 1.33: Symbol for a gold cap

1.16 Push-button

The tutorial kit includes a push-button as shown in the following picture. Both connections are broken out twice. Thus, this variant is especially suited for usage on a breadboard. An electrical connection is established as long as the button is depressed. When you disengage the button, the connection is broken. Make sure to connect the four legs of the push-button as shown in the picture. Because of the inner connections the push-button will not work properly if inserted the wrong way round.

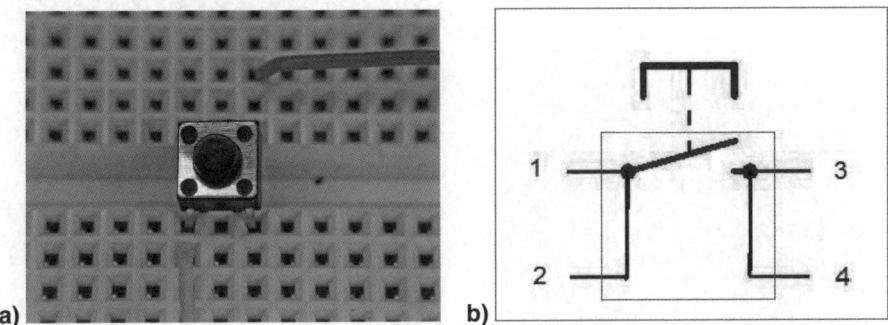

a)   b)

Figure 1.34: (a) Push-button on the breadboard (b) Layout of the connections

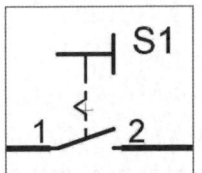

Figure 1.35: Symbol for a push-button

1.17 Battery clip

The battery clip is used to connect a 9V battery to the breadboard. The connection leads (red/black) consist of flexible wire and are tin-plated so that you can insert them without difficulty into the contacts of the breadboard. It is recommended to connect the minus pole (black wire) to the lower bar of the breadboard.

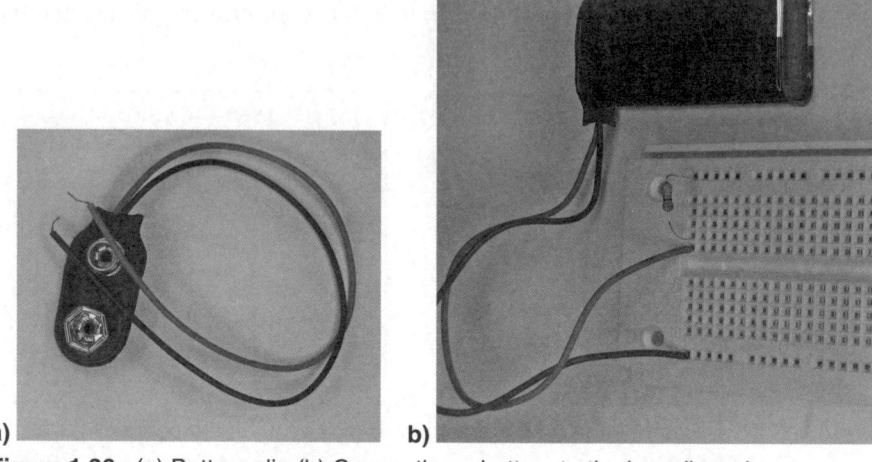

a)   b)

Figure 1.36: (a) Battery clip (b) Connecting a battery to the breadboard

1.18 Cables

When you use the red and black cables with attached alligator clips you can connect individual parts quickly and easily without a soldering iron or a screwdriver. It is sensible to use the red cable for the plus pole and the black one for the minus pole.

Figure 1.37: Cables with alligator clips

1.19   Jumper wire

The tutorial kit contains some jumper wire. Remove approx. 8 mm of the insulation at the ends. Then you can insert the ends directly into the contacts of the breadboard. In order to ease the insertion you can pinch off the ends with a wire cutter in an oblique fashion. You can cut off various lengths of wire as necessary und strip off the insulation at the ends. You can then use these pieces of wire to interconnect the contact rows of the breadboard e. g. to merge the connections of electronic components. Once you have made such a wire jumper, you can reuse it at any time.

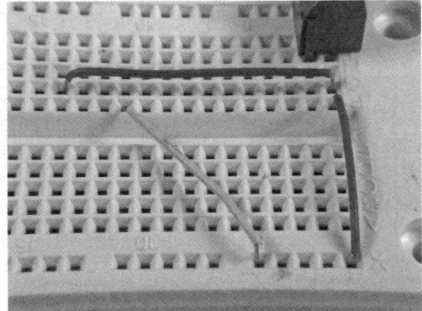

Figure 1.38: Possible usages for jumper wire

1.20   Cut-out templates

The tutorial kit contains two cardboard cut-out templates. Imprinted on them are the following components that you can cut out and assemble as described further down:

o Generator housing for the electrical machine
o Flap wheel, to be used as a water wheel or as a drag-type wind turbine
o Wind rotor blades and adapter disc
o Replacement for flap wheel and adapter disc
o Black cardboard for solar thermal experiments
o White cardboard for solar thermal experiments (back)

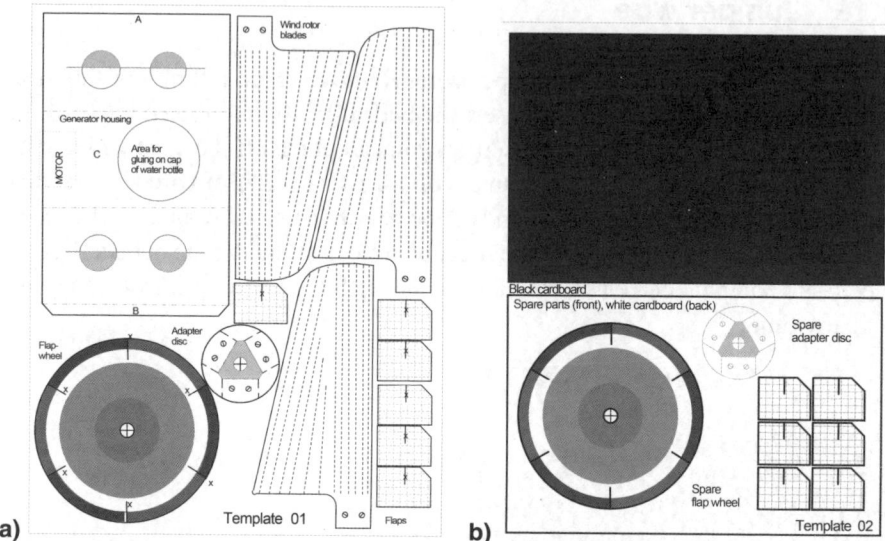

Figure 1.39: Cut-out cardboard: (a) Template 01 (b) Template 02

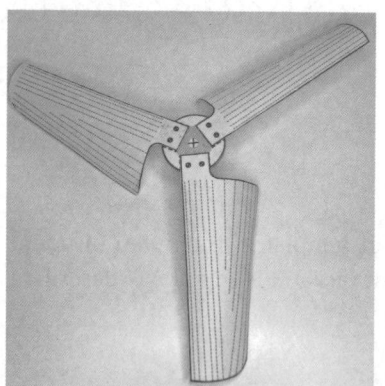

Figure 1.40: Assembled wind rotor

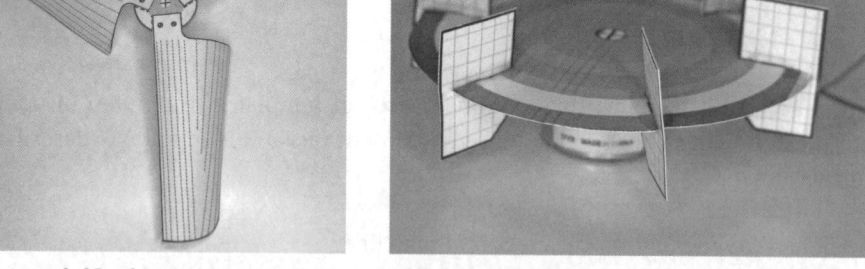

Figure 1.41: Assembled flap wheel

On the white cardboard you will find a copy of the flap wheel and the adapter disc as spare parts. Before cutting out these parts you can use the back of the white cardboard for experiments.

1.21 Disposable syringes

The disposable syringes contained in this tutorial kit will be used for some experiments, e. g. related to solar thermal energy.

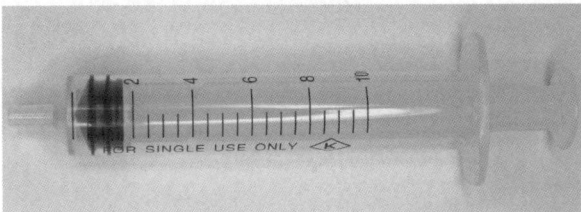

Figure 1.42: Disposable syringes

1.22 Plastic bags

The tutorial kit contains some plastic bags. They are needed for experiments with solar thermal energy.

Figure 1.43: Plastic bag

2 Introduction and Basic Information

When experimenting with natural energy sources, the experimenters should be able to adapt themselves to nature. One of the most important experiences in experimenting with renewable energy is to learn how you can use this energy in cooperation with nature. The tutorial kit covers the following topics:

- Direct use of sun heat (solar thermal energy)
- Generating electrical current out of solar energy (photo voltaics)
- Wind power
- Water power
- Storage technologies

In the experiments, the various forms of energy are made perceptible to you. The experiments guide you from basic knowledge to practical applications and measuring technology to storage technologies.

In addition, there are also other forms of renewable energy:

- Biomass
- Biogas
- Biofuel
- Thermal cells
- Fuel cells
- Tides (tidal power plants)
- Geothermal energy

3 Origin of Renewable Energy

Basically, all the renewable or *regenerative* energy sources (wind, water, bio-mass, solar thermal energy and photo voltaics) derive their energy from the sun. These forms of energy distinguish themselves by immediatley closed circuits, especially concerning the CO_2 balance, and by their decentralised nature that makes them usable for all. This independence from 'industrial' energy providers offers great advantages:

○ Responsible use of the environment
○ Sustainability
○ Protection of precious energy resources
○ Cooperation of man and nature
○ Combination of sources and technical possibilities
○ Social and economical models for the benefit of all

Not counting wind power and water energy, the sun provides for approx. 1000 kWh of energy per square metre (annual average for Germany). This corresponds to the energy content of approx. 100 l fuel oil or 100 cubic metres of natural gas. The quantity of useable energy depends on several factors, e. g. the technical possibilities and the applications. Alignment with the sun and sunshine duration have an essential impact. In order to efficiently use renewable energy, the components of the power plants must be reasonably sized and adjusted to each other.

Due to the rising energy costs, technologies for the use of renewable energy become more and more reasonable. Nature does not bill us! The sooner we begin, the more energy we can harvest and thus save money and relieve the burden of dangerous emissions on our environment.

4 Experiments on Solar Thermal Energy

The enormous energy flow of the sunlight transports vital warmth and light to earth. The greater part of the solar radiation that reaches the earth consists of thermal radiation. Until the energy of the sun reaches the surface of the earth, it has to pass the protective shell of the earth's atmosphere, any natural clouds and the air clouded by pollution. The individual shares of several parts of the spectrum in this radiation are given in the table below. The radiation reaching the earth – called *global radiation* – comprises direct and diffuse radiation.

Wave length (nm)	Ultraviolet 0–380	Visible range 380–780	Infrared 780–2500	Solar constant (total)
Intensity (W/m²)	98	640	618	1353
Share of the total radiation (%)	7	47.3	45.7	100

Table 4.1: Composition of solar radiation. The solar constant in the rightmost column is reduced by factors like the atmosphere and the air pollution so that the maximum intensity (global radiation) that reaches the surface amounts to approx. 1000 W/m².

The average annual global radiation amounts to approx. 920 W/m² in northern Germany and up to 1240 W/m² in the southern parts.

Solar radiation can be used directly as thermal radiation or be converted into other forms of energy like e. g. electrical current.

Thermal radiation is called *solar thermal energy*. It is possible to directly use the solar energy as thermal energy without converting it into another form of energy. There are various possibilities from simple to highly complex technologies.

Photo thermal or solar thermal energy

The thermal radiation of the sun can be absorbed and collected with a solar collector and then provided to domestic uses.

Solar thermal plants can be used for heating tap water, cooking, heating and cooling of rooms.

Figure 4.1: Thermosiphon plant for hot water production by solar energy

Construction and mode of operation of a solar thermal collector differ conceptually from construction and mode of operation of a solar cell (the solar module in the kit). Solar collectors catch and absorb the solar radiation. The heat is transported via a heat carrier (like water, oil or air) and not converted directly into electrical current. Such a collector can only be used for warming up water or supporting the heating in living quarters or to drive a turbine.

Typical and proven applications are systems for heating up water like e. g. simple solar plants on the roof in southern countries. These plants typically consist of a field of collectors and a horizontal or upright storage above them. Since there is no danger of freezing in these countries, these plants can be constructed as single-circuit systems. The tap water circuit and the solar circuit are merged. Furthermore, these solar plants mostly work as thermosiphon systems: warm water expands, gets lighter and floats up, cold water contracts, becomes heavier and sinks down. They do not need any pumps and no additional power supply for agitation.

In colder regions dual-circuit plants working according to the principle of gravity can be used:

○ Solar thermal circuit with anti-freezing agent
○ Hot water circuit with tap water

Today, many buildings are provided with hot water that has been warmed by solar collectors. In the transitional seasons, solar energy can save more than 60 % of the required energy for the production of hot water, and by supporting the heating system, it can save between 30 and 50 % of the energy for heating as well.

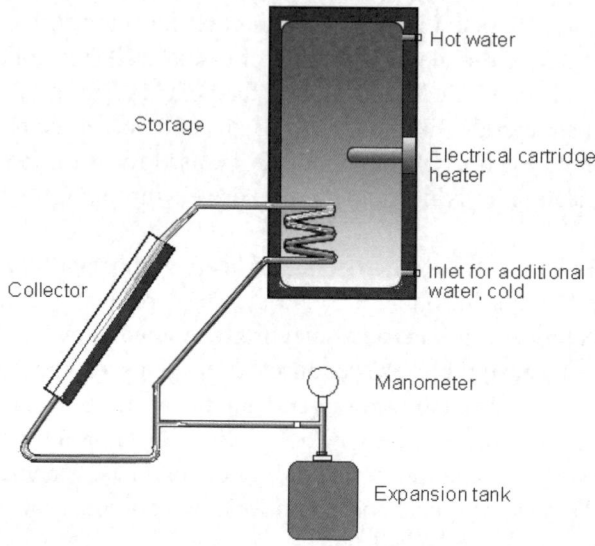

Figure 4.2: Basic structure of a dual-circuit thermosiphon plant: Cold water in the lower part of the storage sinks down to the collector. There it is heated up by the sun, expands and becomes lighter and therefore floats up into the storage and transfers its heat to the storage medium. Then it gets heavier, sinks down into the heat exchanger and picks up heat again.

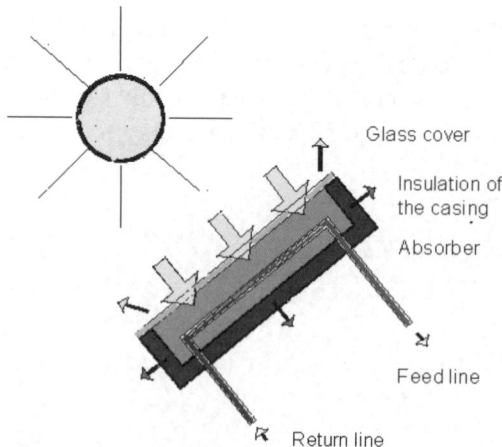

Figure 4.3: Basic structure of a solar collector

4.1 Preparing and using the gauge

Required parts: Breadboard, moving-coil gauge

The tutorial kit contains a moving-coil gauge that you will use in several of the following experiments. Insert the connection leads of the gauge into the contacts of the breadboard – the black wire (minus pole) into the lower bar and the red wire (plus pole) into one of the upper contacts. The ends of the connection leads are tin-plated and can directly be inserted into the breadboard contacts.

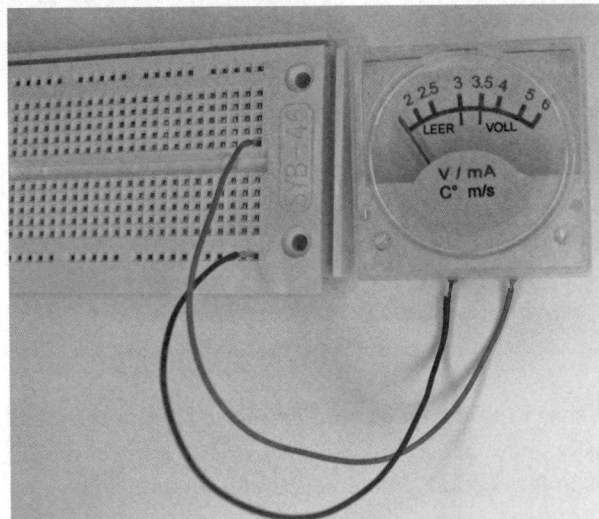

Figure 4.4: Connecting the gauge to the breadboard

> **Note**
> Never connect the moving-coil gauge directly to a battery because this could damage it.

The scale of the moving-coil gauge is divided by seven tick marks (0 to 6) and can be used for various measuring tasks.

4.2   Measuring temperatures

A temperature measuring device is an important and reasonable requirement for the experiments on solar thermal energy. With the components contained in the tutorial kit, you can make your own electronic temperature measuring device for extensive experimentation.

> **Note**
> Each experiment in this chapter builds upon its predecessor. Thus, it is not necessary to dismount the prior setup. Instead, you add, remove or replace individual parts to extend it.

Constructing a temperature measuring device

Required parts for the basic circuit: breadboard, LM358 IC, six 1-kΩ resistors (brown-black-red), one 10-kΩ resistor (brown-black-orange), 10-kΩ trim potentiometer, NTC sensor, moving-coil gauge, battery clip, 9V battery, push-button, two pins

Assemble the temperature measuring device according to the illustrations and the circuit diagram. All resistors but one have a resistance of 1 kΩ. Insert the IC according to the instructions in the section 'Integrated circuit/op-amp'.

Insert the components into the breadboard as shown in the picture. Postpone the connection of the battery clip to the 9V until you are ready to calibrate the circuit. As described below, the circuit will be activated later on by the push-button.

Before you can use the measuring circuit it has to be calibrated so that the needle deflection of the gauge relates to the temperature. For the experiments in this book a rough calibration will suffice because we are primarily interested in the trend, i. e. we want to know whether the temperature rises or falls or whether it is higher or lower in one area as compared to another. When you have an exact temperature measuring facility at your disposal, you can of course use it to support the calibration. Otherwise it is sufficient to calibrate the temperature measuring circuit based on an estimated room temperature of 20 °C.

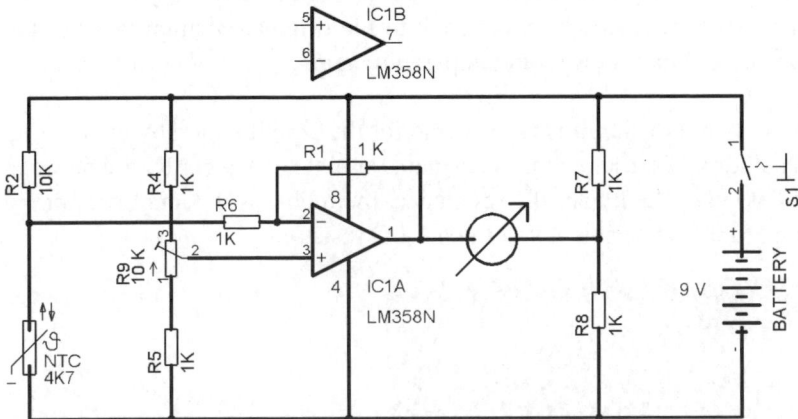

Figure 4.5: Basic circuit

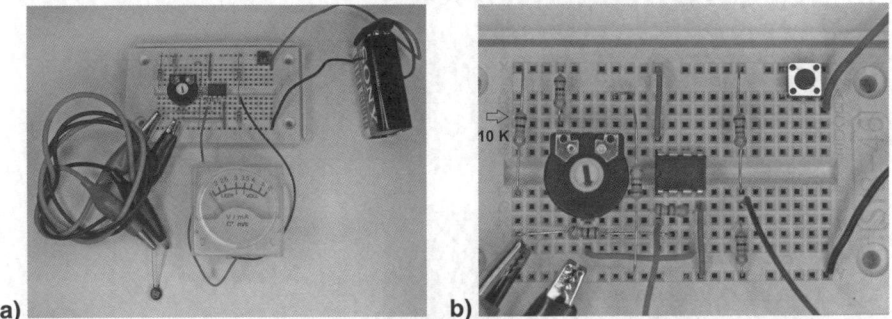

a)   b)
Figure 4.6: (a) Setup of the experiment (b) setup on the breadboard

After having assembled the circuit, you set the slot in the trim potentiometer to the middle position. Now you connect the battery. Then you adjust the deflection of the gauge via the potentiometer. At room temperature of approx. 20 °C you turn the potentiometer by means of a screwdriver until the needle points to the imprinted '2' mark on the scale. At the same time you have to keep the push-button depressed in order to activate the circuit. If the needle of the gauge is not deflected, you will have to release the button and check the proper wiring of the circuit.

When you are done with the calibration, hold the NTC between your thumb and your index finger and depress the button. The sensor gets warmer, and thus the needle moves to the right to show the rising temperature. When you let go the sensor, the needle will slowly move back to the original position because the NTC cools off and measures room temperature again.

The resistor *R1* in the circuit is responsible for the amplification by the op-amp and the resolution of the temperature display. Possible values for R1 are between 1 kΩ and 300 kΩ. The higher the resistance, the higher is the amplification by the op-amp. In this experiment you use a 1-kΩ resistor.

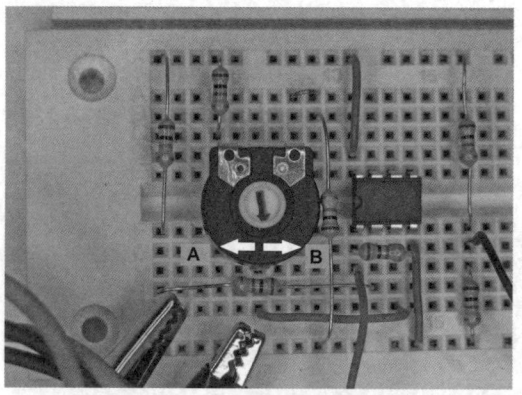

Figure 4.7: Adjusting the needle of the gauge by turning the potentiometer clockwise (A) or anticlockwise (B)

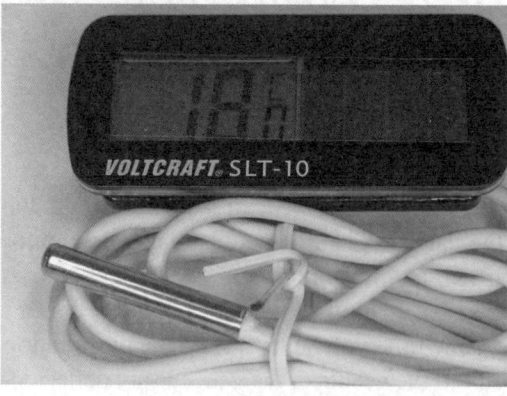

Figure 4.8: Using another device as reference for the temperature display (here a Voltcraft SLT-10 thermometer)

4.3 Experiments with dark and light surfaces

Is it true that dark surfaces heat up more rapidly than light ones? In order to check this we assemble the following experimental setup.

Cut out the two rectangles labelled 'black cardboard' and 'white cardboard', respectively, along the border line. The white cardboard rectangle includes further cut-out templates for other experiments. For this experiment you place it with the completely white back towards the sun.

First, you attach one NTC sensor each to the backs of both cardboards (as close to the centre as possible). Use some adhesive tape and make sure that the connection leads of the sensors remain accessible.

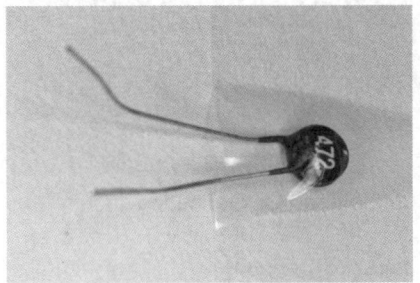

Figure 4.9: Attach the NTCs with adhesive tape to the backs of the cardboard rectangles.

Now place the cardboards for 10 to 20 minutes into direct sunlight. In regular intervals, say five minutes, connect the NTCs via the alligator clips and cables to the circuit and depress the push-button in order to measure the temperature. It does not matter which of the connection leads you connect with which cable.

Now you can read off the values displayed on the moving-coil gauge and record them together with the time specification in a table. You can even draw a diagram based on these values manually or on your computer.

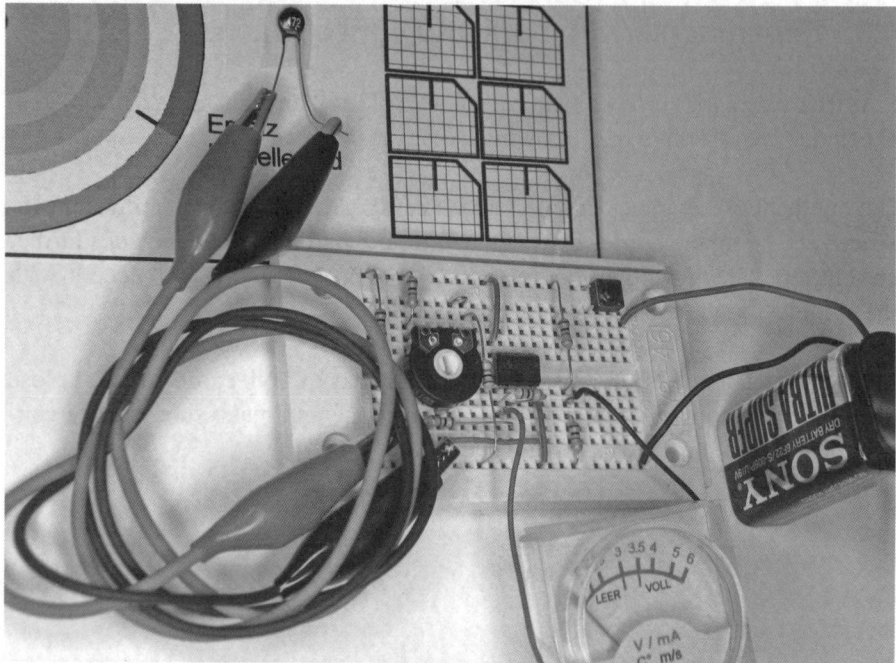

Figure 4.10: Connecting the NTC sensor to the temperature measuring circuit

When you have performed the experiment and evaluated the measured values, you will see that the dark surface warmed up more rapidly and more intensely.

Absorption of solar energy

Absorption means that something is absorbed or 'sucked up'. In order to efficiently use the solar radiation, it should be absorbed as completely as possible and reflected as little as possible. As you have seen in the preceding experiment, the black cardboard heated up more rapidly than the white one. That way it absorbed more of the solar radiation in the same time period.

The reason for this is the greater absorptive capacity of the colour black. The physical explanation: black surfaces can gather more heat because the molecules oscillate faster.

4.4   Greenhouse effect and solar energy

Another effect that can be used in harvesting solar energy is the technical greenhouse effect. This must not be confused with the greenhouse effect in the atmosphere that is caused by environmental pollution (emissions).

This effect is used in greenhouses and hotbeds to facilitate better growing conditions for the plants (e. g. in the beginning of spring). The glass is nearly completely permeable for short-wave solar radiation. Depending on the type of glass, only a small part of the radiation is reflected or absorbed on the glass surface.

The more the incidence angle coincides with the perpendicular, the more radiation penetrates the glass. When the short-wave radiation hits a surface after passing through the glass, some part of it is reflected, but the greater part is absorbed. As you have seen in the preceding experiment, it depends on the surface (colour) how much of the radiation is absorbed.

By the absorption of the solar radiation, the temperature of the surfaces behind the glass walls rise so that these surfaces emit thermal radiation themselves. However, this is long-wave radiation (infrared) that cannot freely penetrate the glass. Because of this 'heat trap', the temperature in the region behind the glass rises until the excess is dissipated by a temperature-dependent heat transfer through the glass (transmission heat loss). For solar collectors it is therefore necessary to use glass panels and frames with an adequate thermal resistance value in order to use as much of the trapped heat as possible.

The cardboards with the attached temperature sensors from the preceding experiment will now be used for another experiment.

Required parts: bottom part of the tutorial kit's box, black cardboard with temperature sensor, cables with crocodile clips, temperature measuring device, transparent cover for the box (wrapping film or glass plate)

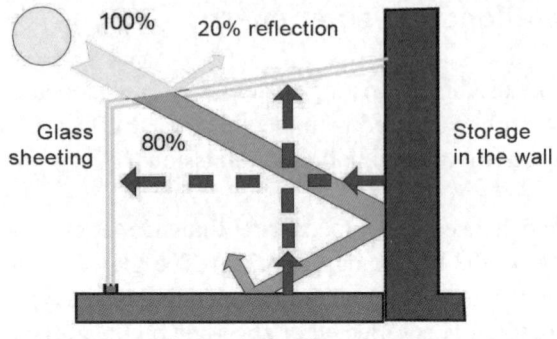

Figure 4.11: Principle of the greenhouse effect: Short-wave solar radiation is converted into long-wave thermal radiation.

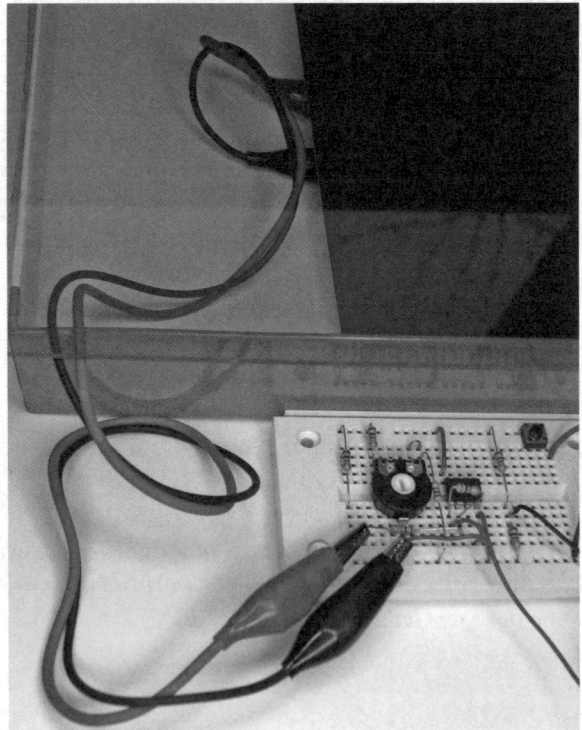

Figure 4.12: Setup of the collector and the measuring circuit

4.5 Solar collector for using the solar thermal energy

The mode of operation of a solar collector is the combination of the above mentioned means to trap and absorb the solar radiation as efficiently as possible.

In order to use the greenhouse effect technically you need a transparent cover, and to absorb the trapped solar radiation an absorber is required.

The mode of operation; solar radiation passes through the glass, the short-wave radiation strikes the absorber and is converted into long-wave thermal radiation. For solar collectors it is important to use a special type of glass that reflects as little of the solar radiation as possible and lets most of it pass towards the absorber. The glass and the insulation of frame and floor contribute to minimising any heat losses. Inside the box, the radiation is sucked up by the absorber and passed to the thermal conduction medium (air, water with anti-freezing agent, or oil). The flat black coating of the absorber yields a high degree of efficiency.

Figure 4.13: Typical flat plate collector mounted on a roof

Performance and degree of efficiency of the collector depend on a stable, tension-free and leakproof frame. Steam arising inside the collector must have a means to escape the casing because otherwise, the transparent cover will cloud over, reducing the output.

The material used for the collector, e. g. for the absorber and the insulation, must be temperature-resistant. Flat plate collectors can heat up to 200 °C in extreme cases (e. g. at a standstill). In addition, the transparent cover must be weatherproof, e. g. resistant against snow, ice and hail.

4.6   Experiments with the solar collector

Required parts: bottom part of the tutorial kit's box, black cardboard, two syringes, one plastic bag, wrapping film or glass plate

Use the bottom part of the tutorial kit's box with the open side up. Place the black cardboard on the bottom, fill one syringe with water, put it into a plastic bag and place it in the box. Now cover the opening of the box with wrapping foil or a glass plate and place the setup for approx. 1 hour in the sun. Fill the other syringe with water und place it in the sun in the vicinity of the box. Afterwards feel the temperature difference with your hands and compare the temperature of the water in both syringes.

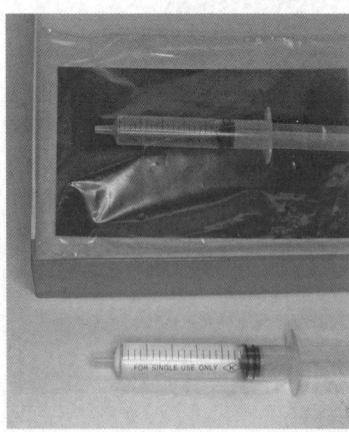

Figure 4.14: Setup with the bottom part of the kit's box

Principal setup for a series of further experiments

Attach an NTC sensor to one syringe using adhesive tape. Connect the sensor via the alligator clips and cables with the temperature measuring circuit. Assemble one of the following experimental setups after the other. Position each in a place with direct sunshine and observe it for 15 to 30 minutes, reading off the temperature values on the moving-coil gauge. Optionally you can insert the second NTC sensor directly into the breadboard in order to determine the 'environmental temperature' for comparison.

Series of experiments with an NTC sensor

a) A syringe, placed directly in the sun
b) A syringe as before, but placed inside a plastic bag
c) A syringe as before, but placed inside a plastic bag together with black cardboard
d) The setup of (c) placed in the box of the tutorial kit and aligned with the sun
e) The setup of (d), but this time covered with wrapping film or a glass plate

Now you can monitor setup (e) for a longer time period and determine to which value the temperature rises. You can also experiment with different alignments to the sun. When everything works optimally, the needle of the moving-coil gauge will be deflected up to the right stop. When this happens, you can assume that the temperature of the water inside the syringe amounts to 50–60 °C.

Using the second NTC, you can determine the normal environmental temperature of about 18–20 °C for comparison. This means that you achieved a temperature difference of 20–40 °C with this simple setup. Depending on the season in which you perform the experiment, the water inside the collector can become hot enough to make tea.

Now place the whole setup in the shadow and watch the temperature going down.

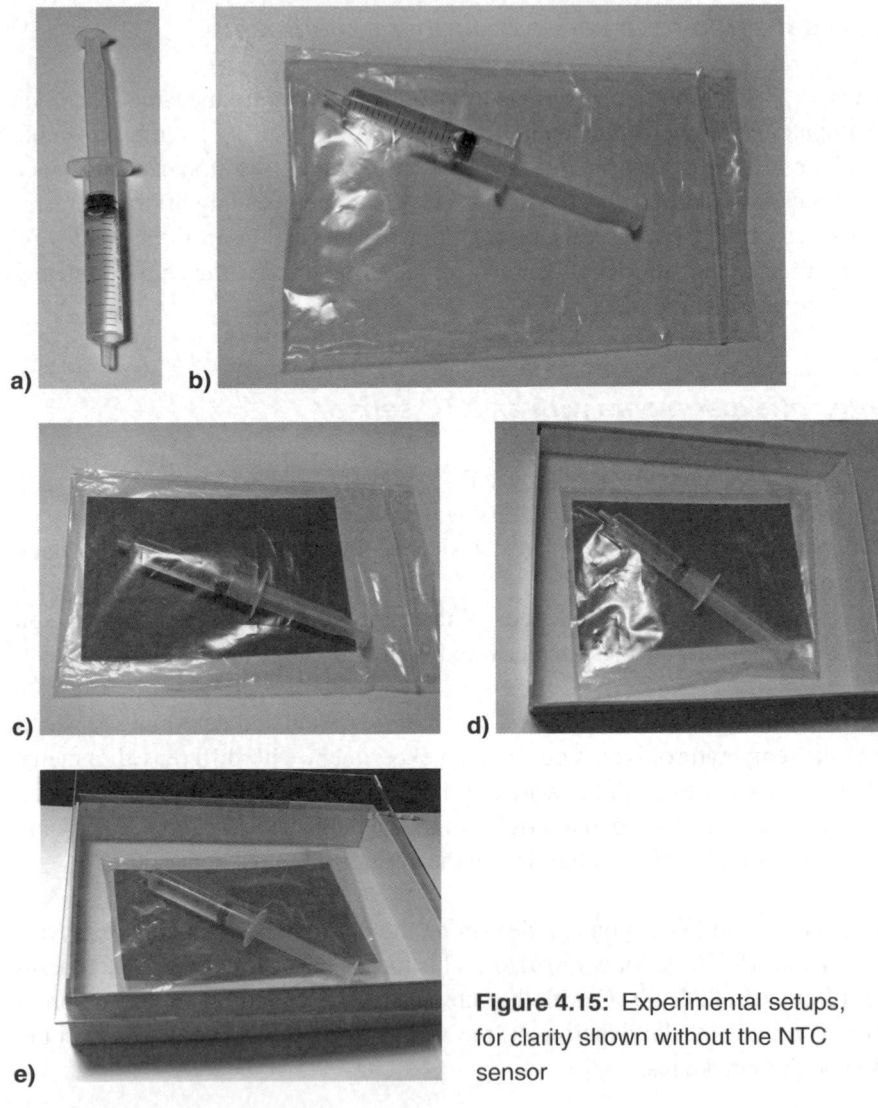

Figure 4.15: Experimental setups, for clarity shown without the NTC sensor

For practical uses it is important to store the captured solar energy for a time when it is needed but the sun does not shine. The are several technical options for this. When you store the energy directly in the collector, the device is called a storage collector. It is also possible to use a separate storage device that can be placed in the boiler room. In this case, further technical components are needed, e. g. a solar pump and a control unit, so that the thermal energy can be transferred from the collector to the storage device.

4.7 Solar thermal energy plant and control electronics

A solar thermal energy plant must have a control unit that decides whether the medium inside the collector (e. g. water) is hot enough to transfer it to the storage. This unit is usually called a *solar regulator*. In the most basic variant there is one temperature sensor in the collector and another one in the storage. When the collector temperature is higher than the storage temperature, the regulator tells the solar pump to transfer the hot medium in the collector to the storage.

Required parts: modified basic circuit of the temperature measuring device with an additional 100-Ω resistor (brown-black-brown), 2.2-kΩ resistor (red-red-red) as a replacement for R1)

Modify the basic circuit according to the circuit diagram and the picture. The second NTC is inserted directly into the breadboard. The push-button is not used this time. Therefore you have to insert the red connection wire of the battery clip into the upper bar.

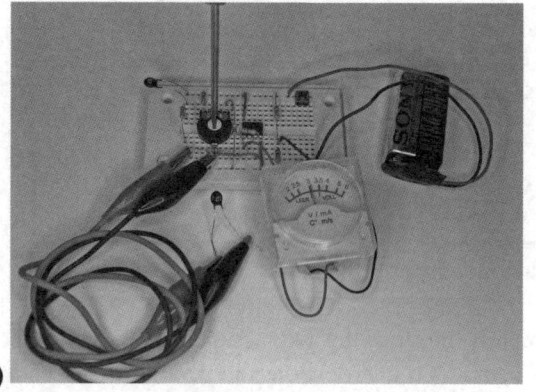 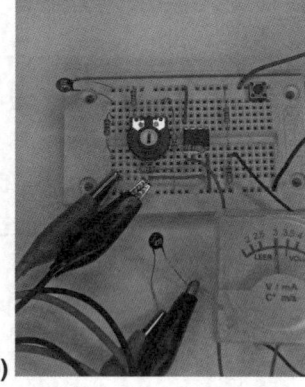

a)　　　　　　　　　　　　　　　　　　　　　　b)

Figure 4.16: (a) Setup for temperature difference measuring with the moving-coil gauge (b) Connection of the second NTC temperature sensor

When you have modified the circuit, you connect the battery and adjust the needle of the moving-coil instrument so that it is approximately in the middle of the scale (at mark '3').

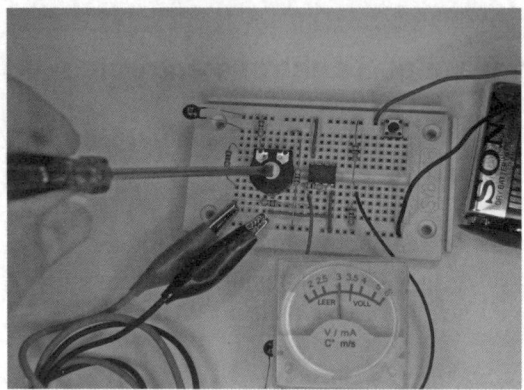

Figure 4.17: Adjusting the measuring circuit with a screwdriver

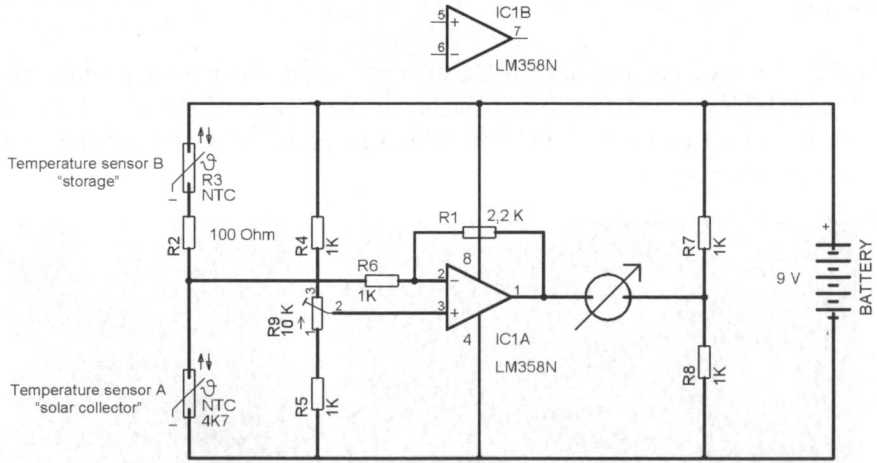

Figure 4.18: Circuit diagram of the measuring circuit

Now you experiment by alternately holding the two temperature sensors between your thumb and index finger. You can observe the needle of the gauge moving to the right and to the left.

Series of experiments

Touch sensor A. The needle of the gauge moves to the right. This means a rising temperature at the sensor and thus a greater temperature difference.

Now touch sensor B. The needle goes to the left because the temperature difference between the two sensors is reduced.

In yet another experiment you use this electronic circuit in combination with the solar collector made out of the bottom part of the box.

Attach the NTCs sensors to the syringes using adhesive tape. The setup is similar to that in Fig. 4.15(e). Fill syringe B with warm water, connect NTC sensor A via the alligator-clip cable with the breadboard, and insert the leads of NTC sensor B directly into the breadboard. Adjust the trim potentiometer so that the needle points to mark '3' on the scale. Connect the battery and align the collector setup with the sun.

a)

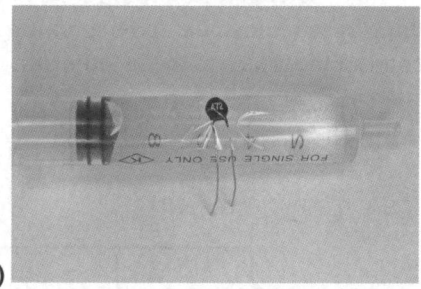

b)

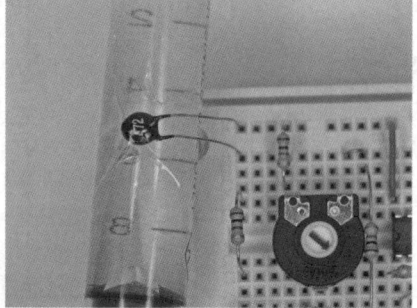

c)

Figure 4.19: Experimental setup with solar collector: (a) syringe A inside the collector, connected via the alligator-clip cables; (b) syringe B representing the storage device in the boiler room; (c) attaching sensor B to the breadboard (A and B see also Fig. 4.20)

Observations

At the beginning of the experiment, the water in the storage is warmer than the water in the collector. Then the sun heats up the water in the collector, while the water in the storage cools down. The needle of the moving-coil gauge moves to the right. In this way it shows that the water in the collector is now warmer so that it is worthwhile to transfer it to the storage.

The next step is to use the electronic circuit to control a motor that is activated according to the temperature difference between the two NTC sensors.

4.8   Solar regulator with pump control

Required parts: basic temperature measuring circuit as in the preceding experiment, 2N3904 transistor, 1-kΩ resistor (brown-black-red), motor

Modify the circuit according to the circuit diagram and the pictures. The main thing is to remove the moving-coil gauge and to insert the transistor. You also connect the motor to the breadboard.

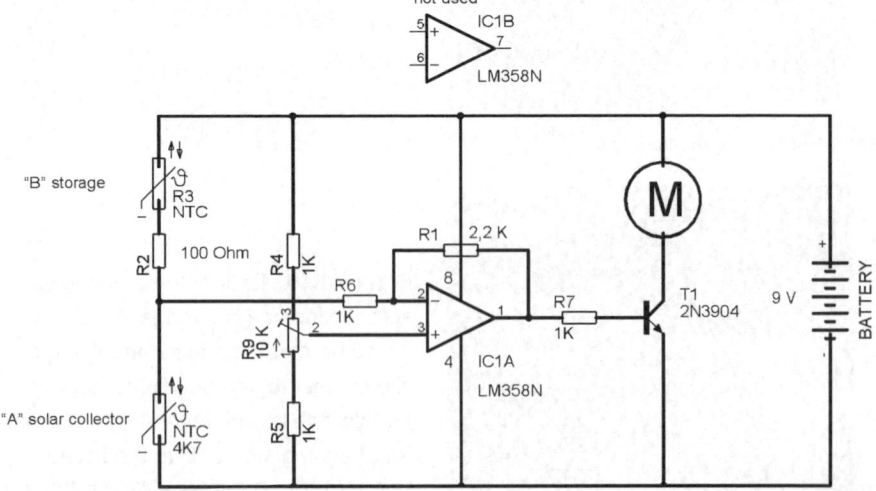

Figure 4.20: Circuit diagram for the experimental setup with the motor

After you have attached the battery clip to the battery, you adjust the potentiometer until the motor shaft begins to rotate. Then you turn the potentiometer back until the shaft stands still. Now repeat the experiment (described above) with the two NTC sensors alternately warmed up between your thumb and index finger.

When you warm up sensor A the motor shaft begins to turn. Warming up sensor B stops the movement.

You can increase the sensitivity of the circuit by replacing the 2.2-kΩ resistor R1 (red-red-red) by a 10-kΩ resistor (brown-black-orange).

In the next step you will extend the circuit in order to control a 'solar pump' (just conceptually!). The pump is to be activated when the collector temperature is higher than the storage temperature.

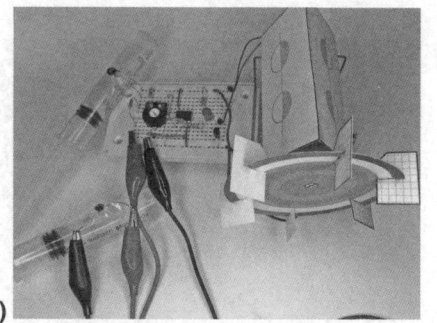 

a)   b)

Figure 4.21: (a) Setup with 'solar pump' (b) setup on the breadboard (role of the LED see below)

Required parts: basic temperature measuring circuit as in the preceding experiment, motor with flap wheel

Instead of a solar pump you will use the motor and the flap wheel contained in the tutorial kit. Just imagine that this construction works as a solar pump.

Adjust the potentiometer so that the needle points to mark '3'. Attach the battery and align the collector directly with the sun. When the collector warms up, the motor shaft eventually begins to turn. This happens when the temperature in the collector is higher than the temperature in the storage, at which point the 'pump' shall transfer the warm water into the storage. When the pump is running, you can place the collector in the shadow or cover it up so that its temperature drops. Additionally you can warm up the content of the syringe with the second NTC sensor (that acts as storage). When the collector has a lower temperature than the storage, the 'pump' will be deactivated.

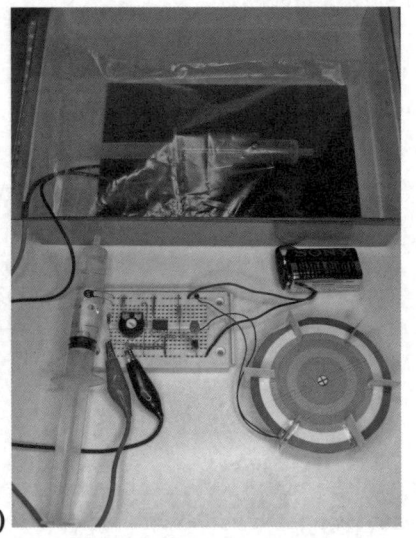
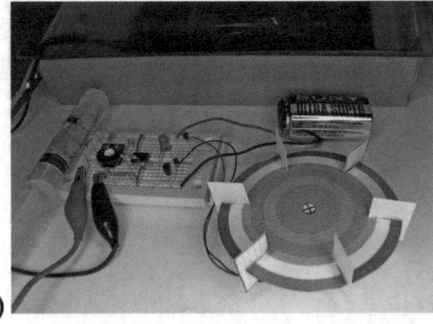

a)   b)

Figure 4.22: (a) Complete setup with collector, circuit and 'pump' (b) detailed view

The principle is as follows: The solar pump is engaged as long as the collector temperature is higher than the storage temperature.

Optionally you can add a red LED with a series resistor of 1 kΩ (brown-black-red) and repeat the preceding experiment. Now the LED lights up when sensor A reads a higher temperature than sensor B.

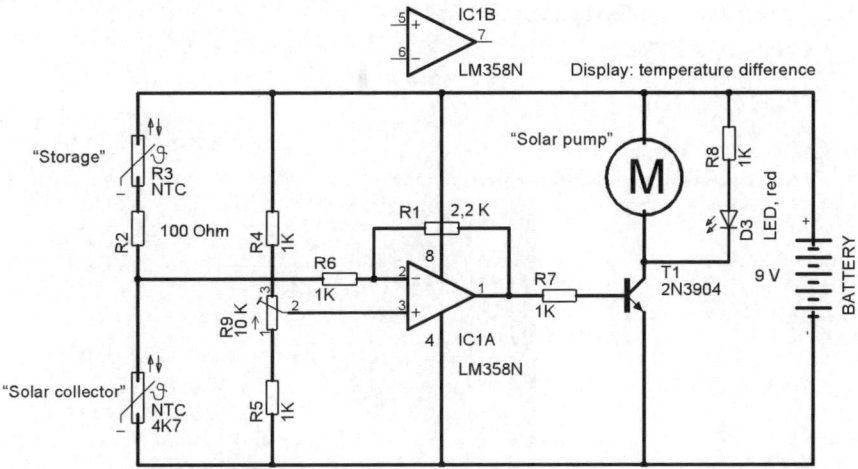

Figure 4.23: Temperature difference circuit with motor and optional red LED

4.9 Using solar thermal energy in the real world

The experimental setups show the mode of operation in a model. The systems used in reality follow the concepts you researched in these experiments. Of course the professional plants perform much better and can easily provide for a great share of the required domestic hot water. Furthermore, there are mature systems that greatly contribute to the heating of a building. These are called supporting solar plants for heating.

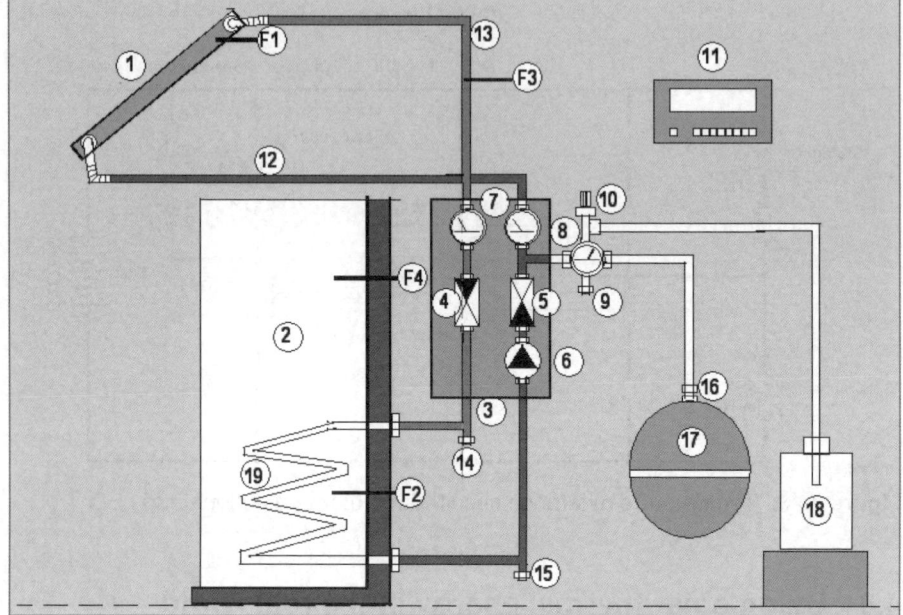

Figure 4.24: Components, pump assembly and regulator of a professional solar thermal plant

Legend to Fig. 4.24

The numbers denote the following components

1. Collector
2. Solar storage
3. Solar station
4. Feed line/bleeder can gravity brake
5. Return line gravity brake
6. Pump
7. Temperature display instruments
8. Manometer
9. Ball valve with supply connection for the solar circuit
10. Safety valve with overflow connection
11. Electronic solar regulator

12. Return line to the collector
13. Feed line from the collector
14. Ball valve with connection for the fill pump (rinsing)
15. Ball valve to drain the solar circuit with connection for the fill pump (rinsing)
16. Automatic block valve (allows unfastening of the expansion tank without draining the solar circuit)
17. Expansion tank
18. Collection receptacle for fluids coming from the relief valve
19. Heat exchanger in the solar storage

Temperature sensors:

F1 = collector

F2 = storage, lower part

F3 = feed line of the solar circuit (warm side)

F4 = storage, upper part (request for reheating)

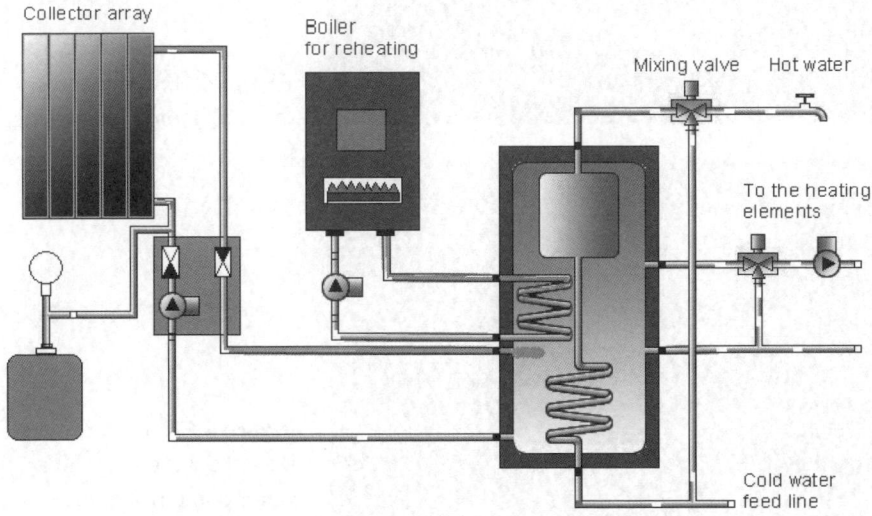

Figure 4.25: Conceptual drawing of a supporting solar plant for heating

In addition, there are many further applications for the practical use of solar thermal energy in everyday life.

Examples of applications for solar thermal energy:
- Cooking and baking with solar energy
- Drying with solar energy
- Melting metal
- Direct thermal decomposition of water into hydrogen and oxygen
- Direct conversion into mechanical energy (e. g. Stirling motor)

Figure 4.26: Using solar thermal energy for cooking

Figure 4.27: Direct conversion of solar energy for a water pump using a low-temperature Stirling motor (Sunpulse)

4.10 Functional principle of a Stirling motor

Solar thermal energy can not only be used to heat up a medium like e. g. air, water or oil, but also to drive a motor, e. g. a Stirling or hot-air motor. In this way, the solar thermal energy is converted into mechanical energy. This mechanical energy can be used to drive a generator and thus to provide electrical energy, or to drive water or air pumps. This means: the more intensely the sun heats up our environment and the hotter it gets for us, the more energy we have at our disposal for driving cooling devices!

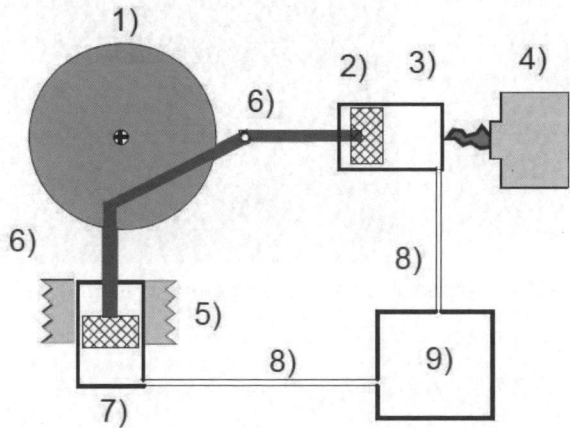

Figure 4.28: Conceptual drawing of a Stirling alpha motor: (1) flywheel (3) displacement body (4) arbitrary heat source (5) colling (7) working piston (9) regenerator

Required parts: A smooth-running syringe, dark liquid like e. g. black ink, a nail to seal the syringe

First test the syringes and use the one with the smoother run of the plunger. Fill the syringe with a dark liquid, e. g. black tea, soy sauce, black ink or the black colour of a felt tip pen dissolved in water. Just fill half of the barrel of the syringe (up to mark '5') and seal the end with a nail or the like. It has to be absolutely leakproof.

Now place the syringe on a piece of black cardboard inside a plastic bag.

Place the setup into direct sunlight and observe what happens.

Warm water or air expand. When the liquid in the syringe is heated up enough, it also begins to expand and presses the plunger out. Now take the syringe out of the plastic bag und cool it down under running cold water or in a freezer. The plunger now moves back into the original position.

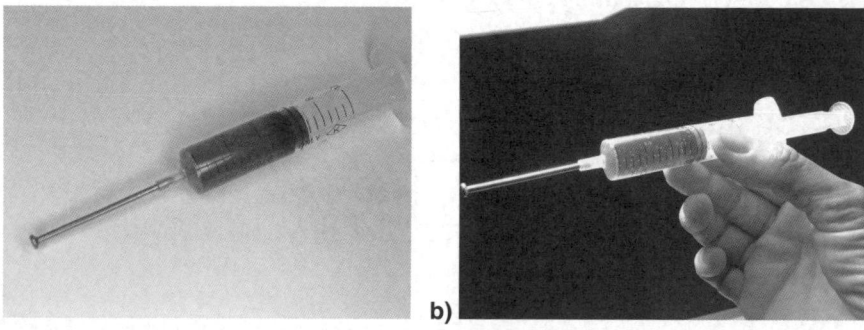

a)   b)

Figure 4.29: Experiment on the Stirling principle using a syringe (a) preparing the syringe (b) using a heat source

Note

These operating sequences are optimised in professional Stirling motors.

A Stirling motor converts thermal energy into mechanical work. The interesting thing is that thermal energy can be fed into the motor from outside. In contrast to a gasoline or diesel motor, a Stirling motor is thus not dependent on the internal combustion of a special fuel but can use arbitrary heat sources. Amongst others, this can be solar energy, waste heat from various technical processes, combustion heat produced by biogas, dump gas or other solid or fluid regenerative fuels. Another benefit is that the combustion of these fuels happens with a high degree of efficiency and is therefore environmentally friendly.

Mode of operation of a Stirling motor

At the first glance, the mode of operation of a Stirling motor appears to be unusual because we are accustomed to the working principle of rapid consecutive explosion as it is used in combustion engines. The Stirling motor follows a completely different principle. It has a warm side and a cold side. A displacement plunger (not sealed against the cylinder) transfers a gaseous medium (e.g. air) to and fro between the two sides. When the medium is on the warm side, it expands, when it is on the cold side, it contracts. This expansion and contraction is converted into mechanical energy by a second cylinder.

Fig. 4.30 shows a simple illustrative model of a Stirling motor. It consists of a glass cylinder with a styrofoam displacement plunger inside and covered by a membrane cut out of a balloon that acts as the working cylinder. The whole cylinder is now placed upon a tea candle. The displacement plunger inside the cylinder can be moved from the outside with a magnet in order to directly show the working principle of a Stirling motor.

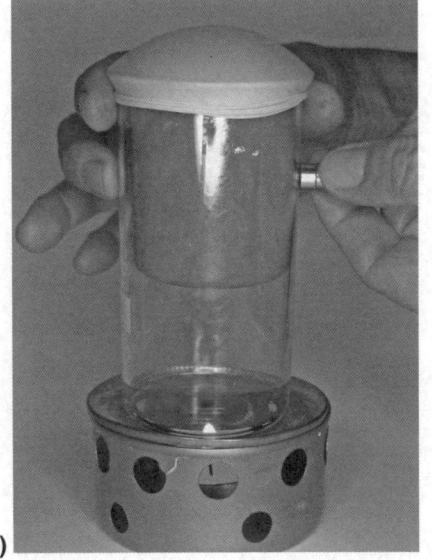

a)   b)

Figure 4.30: (a) When the air is in the warm half of the cylinder, the membrane rises. (b) When the air is in the cold half, the membrane sinks down.

5 Photo voltaics – producing electric current out of sunlight

Photo voltaics is the technical term for the direct conversion of sunlight into electric energy. In this process the solar energy is turned into electric current by means of solar cells or solar modules. This electric current can then be fed into the public network (parallel mains operation) or used directly in the respective household (off-grid system). There is also a number of devices like measuring instruments, illuminated information signs, solar torches etc. that gain their power by solar energy and store it in accumulators.

The tutorial kit contains a solar module for experiments with photo voltaics.

> **Note**
>
> This experiment works even with low light (clouded sky), though the effect is much clearer in bright sunlight.

Place the solar module under a light source. Attach the red and black cables with the alligator clips to the appropriate connection wires of the solar module: the red clip to the red wire of the solar module (plus pole), the black clip to the black one (minus pole). Connect the alligator clips at the other end of the cables with a red LED. Make sure that you attach the red 'plus' clip to the longer connection lead of the LED. Depending on the intensity of the light source, the LED shines more or less brightly. When the LED does not light up, there is either not enough 'light energy' or the LED is attached the wrong way round. If the LED flashes, you have chosen the flashing LED.

Important!

Normally you should never use an LED without a series resistor because this would destroy it. In the experiment described above, you can operate the LED exceptionally without resistor because the experiment does not last very long and the solar module is very small.

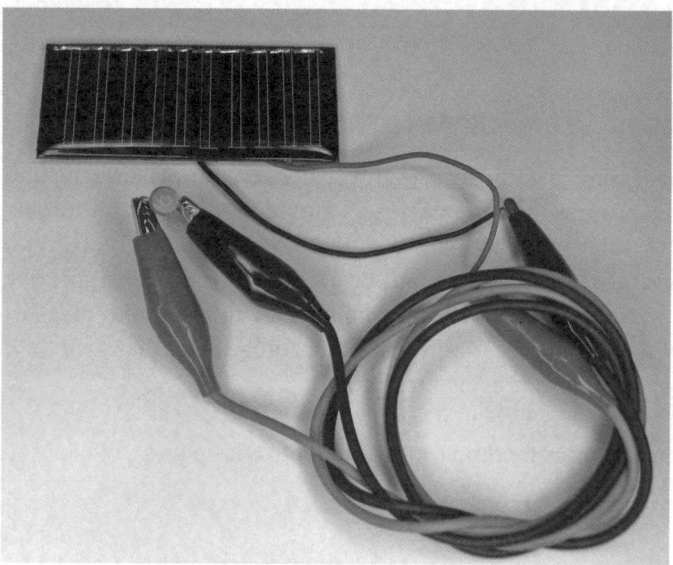

Figure 5.1: Simple functional test with a red LED

You can carry out this experiment with various other light sources, e. g. with:
- direct sunlight
- a halogen lamp
- a light bulb
- a torch
- an energy-saving lamp
- a fluorescent tube
- an LED torch

This shows which light sources are optimally suited for supplying the solar module.

5.1   Connecting the solar module to the breadboard

Required parts: solar module, breadboard, two pins

On the back of the solar module there are connections with soldered-on wires. The module produces direct current. Therefore there is a plus and a minus pole just as it is the case with a battery. Connect the black and the red wire to the breadboard. It is recommended to insert the black connector into the lower bar and the red one into the upper bar. The solar module can stay attached for all of the following experiments.

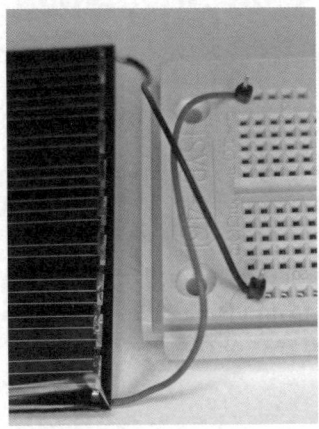

Figure 5.2: It is possible to insert the connection wires of the solar module directly into the bread-board, but the pins can tighten the connection

Place the solar module in such a way that it is illuminated by a sufficiently bright light source.

5.2   Connecting solar cells in series

The solar module contained in this tutorial kit consists of nine or ten individual solar cells with a voltage of 0.5 V each. The cells are connected in series so that the off-load voltage of the module amounts to 4.5–5 V. If you extracted a single solar cell out of the module, it would have a voltage of approx. 0.45–0.55 V, depending on the light source. You could not do anything with that, because this voltage is not high enough to drive even an LED, a transistor or a very simple electronic circuit. LEDs begin to show a very faint light when you apply a voltage of at least 1.5 V. Transistors work from 0.6 V onwards (germanium) or 0.9 V (silicon), respectively.

To gain a higher voltage you need to connect several single solar cells in series. This is similar to electronic devices where two or more battery cells are connected in series.

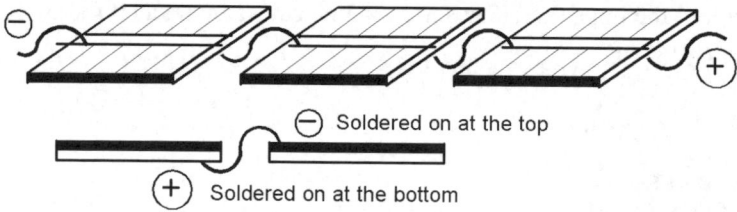

⊖ Soldered on at the top

⊕ Soldered on at the bottom

Figure 5.3: Solar module made out of interconnected crystalline cells; principle of the series connection

Silicon solar cells are connected in series by joining the bottom (backside) of the first cell (plus pole) to the top side of the next one (minus pole) by means of a special conductive flat connector. If you connected two plus or to two minus poles with each other, no current would flow.

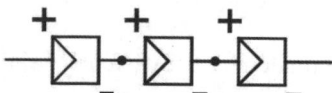

Figure 5.4: Series connection of solar cells in a circuit diagram

The series connection causes the following effects:

○ The voltages of the single cells add up.
○ The short-circuit current is equal to the current of the weakest cell.
○ When one cell is shaded, the power of the complete module is recuded according to the degree of shading.

○ When one cell is partially shaded, the well-lit solar cells feed their current into the shaded one. This one warms up and can even explode in extreme cases.
○ Problems with partial shading arise especially in modules with crystalline cells.

Solar modules used in big PV plants also consist of single crystalline solar cells connected in series. To avoid damages by partial shading, the individual sections of the module are furnished with so-called *bypass diodes* that direct the electric current past the shaded cell.

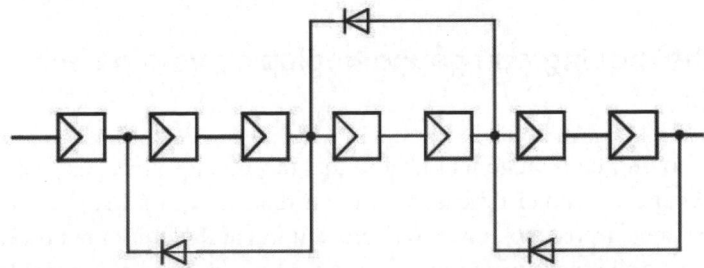

Figure 5.5: Series connection with additional bypass diodes

5.3 Parallel connection

Single solar cells and even whole solar modules can be connected in parallel as well. To this end the minus poles and the plus poles of all cells are connected among themselves.

○ The total voltage of a group of solar cells connected in parallel equals the voltage of a single cell.
○ The total short-circuit current is the sum of the current values of all single cells. When all cells yield the same current, the short-circuit current is simply the product of this value and the number of cells.
○ It is possible to connect cells with different short-circuit current values.
○ When one cell is partially shaded, the well-lit solar cells feed their current into the shaded one. This one warms up and can even explode in extreme cases.

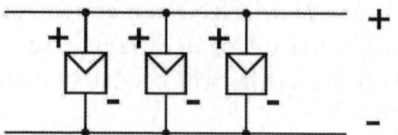

Figure 5.6: Connecting
solar cells in parallel

A parallel connection of solar cells makes sense when you only need a low voltage but high current.

5.4   Using the moving-coil gauge to display voltage and current

You can use the moving-coil gauge included in the tutorial kit to measure various quantities. As the instrument measures the current, it can be used as an amperemeter. For this purpose you have to connect it in parallel with the power source. Depending on the measuring range, you have to add other components in order to adjust the sensitivity of the instruments. For measuring voltages, the gauge can be connected in parallel with the power source, using a series connector.

Measuring currents (short-circuit current)

Required parts: solar module, breadboard, moving-coil gauge, resistors with 100 Ω (brown-black-brown) and 10 Ω (brown-black-black)

> **Note**
> For the following experiments you need light sources of various brightness (shadowed areas and full direct sunlight) for the solar module.

The basic procedure for the experiments is as follows:

a) Measure the short-circuit current (with the moving-coil gauge) of the solar module in shade, full sunlight, half shade and darkness.

b) Try out various measuring ranges to detect low and high currents.
c) Enter the readings into an Excel table or the like and draw a diagram.

Solar module

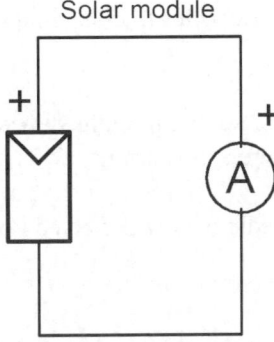

Figure 5.7: Circuit diagram for the wiring of the solar module and the moving-coil gauge

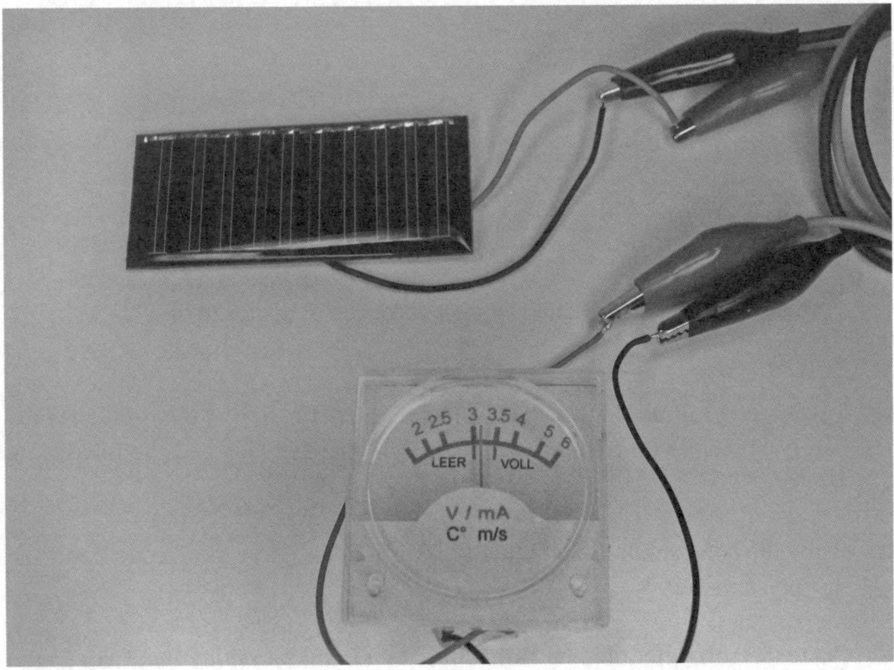

Figure 5.8: Measuring the voltage of the solar module by means of the moving-coil gauge (in the shadow)

In detail, you proceed as follows:

❶ Attach the moving-coil gauge to the solar module that lies in the shade. Make sure you connect the components in the right way (plus to plus, minus to minus). Take a note of the displayed value.

❷ Cover up the solar module with a piece of cardboard so that no light shines on it. The needle of the gauge is not deflected as there is no current.

❸ Place the solar module into direct sunlight. The needle is now deflected up to the stop.

❹ Modify the wiring according to Fig. 5.9.

Solar module

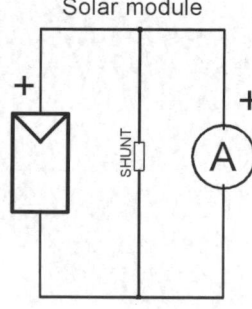

Figure 5.9: Wiring the moving-coil gauge with a 100-Ω resistor. The measuring range is modified to a full deflection at approx. 6 mA.

❺ Take another measurement in the shade. The deflection of the needle is now barely perceptible. With the new wiring you have changed the measuring range.

❻ Place the solar module horizontally into full sunlight and take a note of the displayed value.

❼ Place the solar module into full sunlight and align it optimally. Take a note of the displayed value.

Figure 5.10: If available, use a multimeter set to direct current reading for reference.

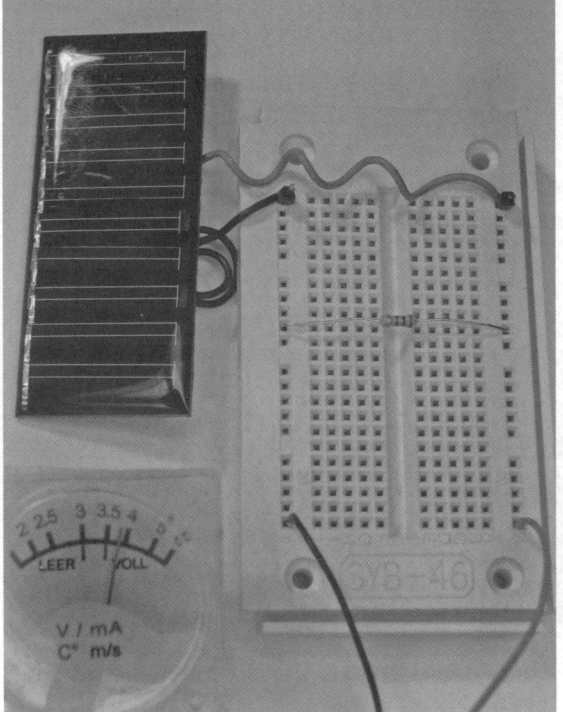

Figure 5.11: Measuring the short-circuit current of the solar module: experimental setup with breadboard and 100-Ω resistor

With the 100-kΩ resistor the value displayed on the scale is the current measured in milliampere. When the needle points to '4', the current amounts to 4 mA. It is possible to measure even higher currents when you direct a part of the current past the gauge. To this end you use a bypass resistor that is usually called a *shunt resistor*. In the following experiment a resistor of 10 Ω is chosen for this task. You can then simply multiply the values imprinted on the scale by 10. Thus, when the needle points to '3', there is a current of 30 mA.

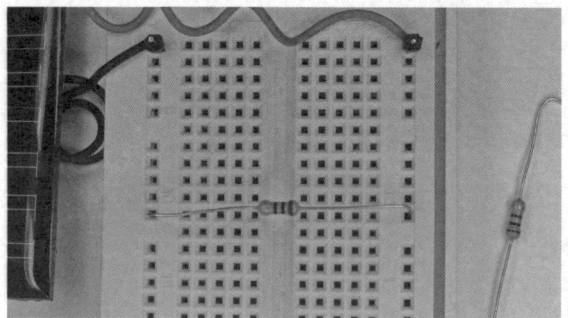

Figure 5.12: Setup with a 10-Ω shunt resistor

If you want, you can enter the values into an Excel table and draw a diagram. It is also possibly to enter them into a simple table on paper and draw the chart manually.

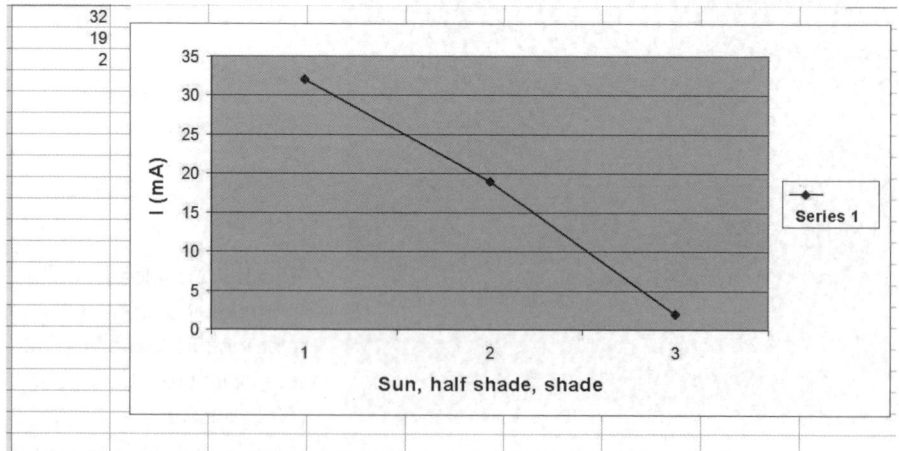

Figure 5.13: Current graph drawn on a computer

Measuring voltages *(off-load voltage)*

Required parts: solar module, breadboard, red LED, 2.2-kΩ resistor (red-red-red), 10-kΩ resistor (brown-black-orange), moving-coil instrument. Additionally, if available: multimeter set to direct current measuring.

> **Note**
> For the following experiments you need light sources of various brightness (shadowed areas and full direct sunlight) for the solar module.

Before you can use the moving-coil gauge to measure the voltage of the solar module, you first have to build a simple circuit that is called a voltage divider. Normally, a voltage divider is exclusively made of resistors. Here you will also use an LED so that you can build the circuit with the components included in the tutorial kit. This has the additional benefit of performing as an operating display that shows whether there is a current flow or not.

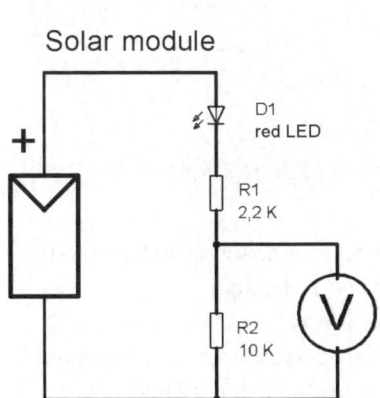

Figure 5.14: Circuit for measuring voltages (voltage divider)

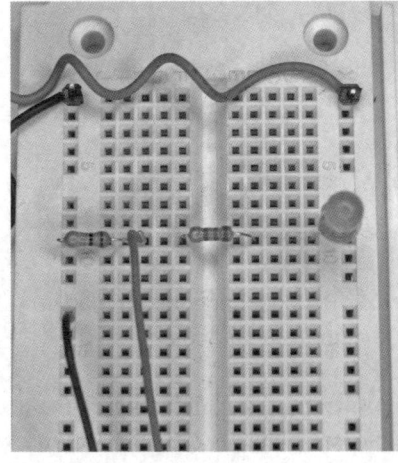

Figure 5.15: Setup of the measuring circuit on the breadboard

The procedure for measuring the voltages is the same as the procedure for measuring currents in the preceding section. The difference is that displayed values of the off-load voltage do not vary significantly with the light intensity.

The basic procedure for the experiments is as follows:
a) Measure the off-load voltage (by means of the moving-coil gauge) of the solar module in shade, full sunlight, half shade and darkness.
b) Enter the readings into an Excel table or the like and draw a diagram.

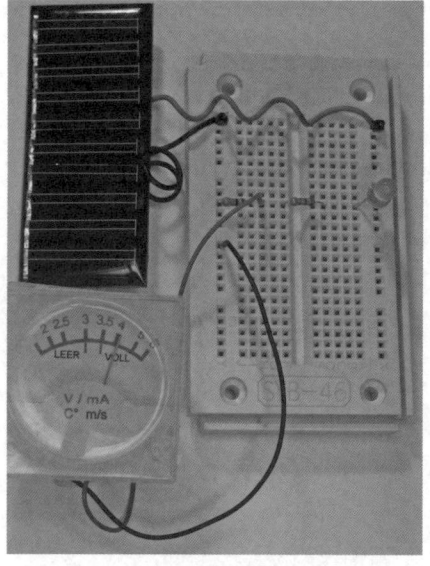

Figure 5.16: Measuring the off-load voltage at the solar module

Figure 5.17: Measuring voltages with the moving-coil gauge

The number on the scale to which the needle points corresponds to the voltage in volts. So when the needle points to mark '4', you measure 4 V. (This is only true in connection with the voltage divider circuit described above!)

In detail, you proceed as follows:

❶ Attach the moving-coil gauge to the solar module that lies in the shade. Make sure you connect the components in the right way (plus to plus, minus to minus). Take a note of the displayed value.

❷ Cover up the solar module with a piece of cardboard so that no light shines on it. There is only a small deflection of the needle. This means there is nearly no voltage.

❸ Place the solar module horizontally into full sunlight and take a note of the displayed value.

❹ Place the solar module into full sunlight and align it optimally. Take a note of the displayed value.

As in the experiment to measure the current, you can enter the values into an Excel table and draw a diagram.

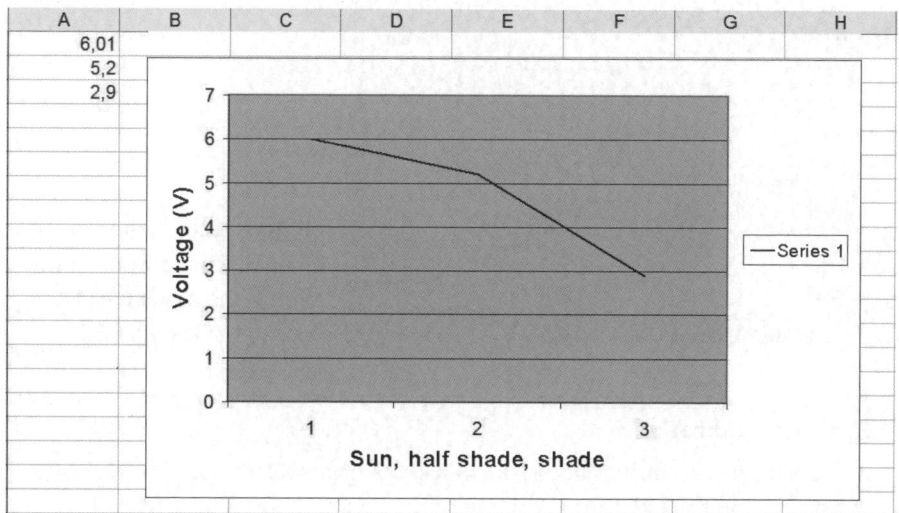

Figure 5.18: Voltage graph drawn on a computer

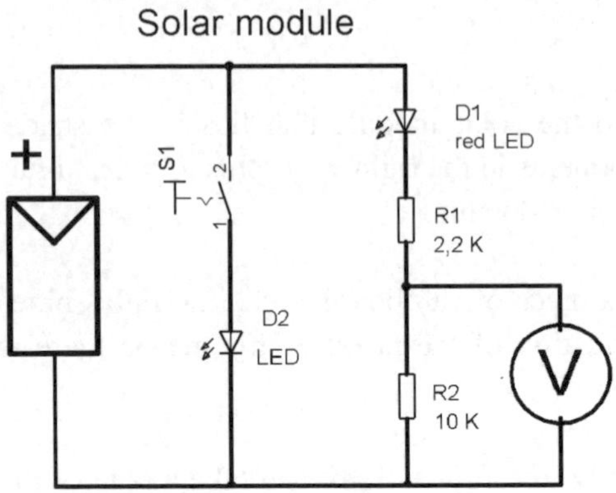

Solar module

Figure 5.19: A consumer that puts a load on the circuit (the orange LED) reduces the off-load voltage of the solar module to the working voltage.

The off-load voltage of the module is higher than the working voltage under load. You check this by connecting the orange LED and the push-button in series to each other and in parallel to the module. When you depress the button the voltage reading decreases. What you now read is the voltage of the solar module in connection with a consumer.

Figure 5.20: Setup for measuring the working voltage in the presence of a consumer like an LED

Real-life application

It is useful to determine the off-load voltage of a solar module so that you can adjust the final charging voltage of an accumulator. When the off-load voltage is too low, the accumulator may not be charged completely.

5.5   What happens when the module is shaded?

Required parts: solar module, breadboard, moving-coil gauge, LEDs with 100-Ω series resistor (brown-black-brown)

Note

For the following experiments you need a bright light source (or full direct sunlight) for the solar module.

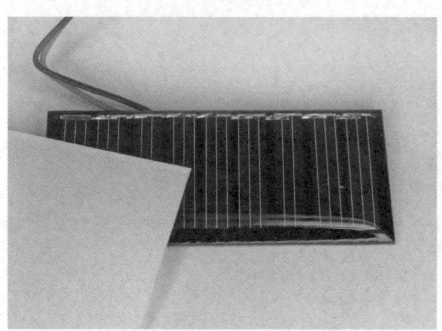

Figure 5.21: What effect does the shading of the module have? (Here: partial shading)

Note

Alternatively, you can perform this and the following experiments with the moving-coil gauge, the LEDs and a multimeter. The flashing LED is suited as well. Just as a reminder: the longer connection lead of an LED is the plus pole.

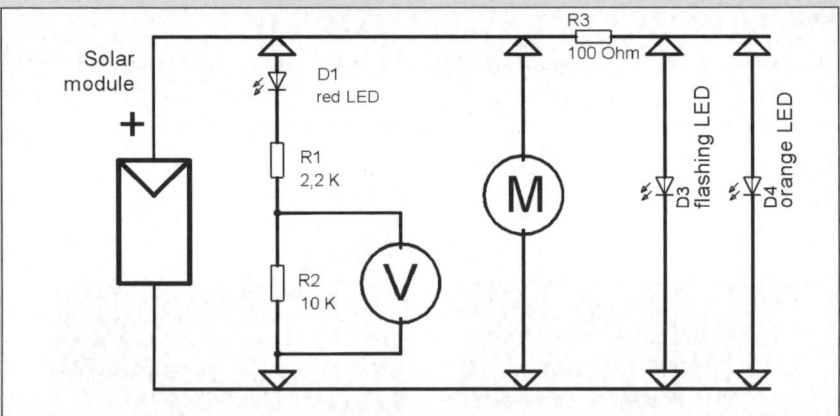

Figure 5.22: Alternative setup with the moving-coil gauge, LEDs with series resistors and motor (as regards the motor, see section 5.8, 'Preparing the solar drive')

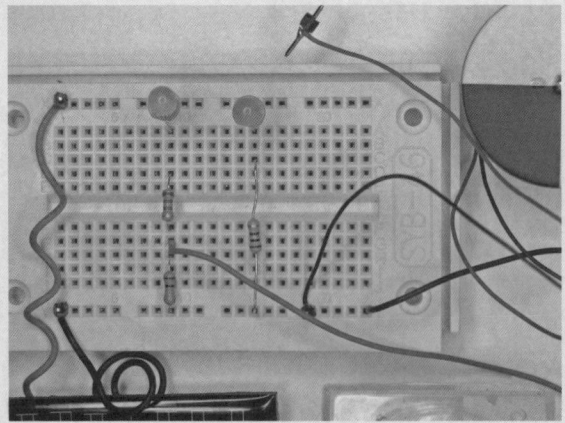

Figure 5.23: Setup on the breadboard

Note (*cont.*)

Perform the experiments outdoors in bright sunshine in order to clearly read the consumption display on the moving-coil gauge. However, in a brightly lit environment the glow of the LED is hard to discern. You can shield the LED with a piece of cardboard against the sunlight.

Now you can perform various other experiments of this type:

Create a light shade by holding an additional glass plate or a dim transparency foil between the light source and the solar module.

Create a dark shade by holding a piece of cardboard or wood between the light source and the solar module.

Shade individual parts of the solar module by placing a piece of cardboard directly on top of the respective area of the module.

Figure 5.24: Shades provided by constructive elements on roof-mounted solar modules

The shading experiments can also be performed with an LED connected to the solar module. What happens to the red, green and flashing LED in light shadow, dark shadow and when covering partial areas of the module?

Note

Shading is a very important topic for big PV plants with crystalline solar modules. Lest a partial shading (e. g. by a leaf) causes an outage of the complete generator, Schottky diodes are used to bypass the current past the shaded cell. Failures in these bypass diodes may cause a *hot spot* that can damage the solar cells.

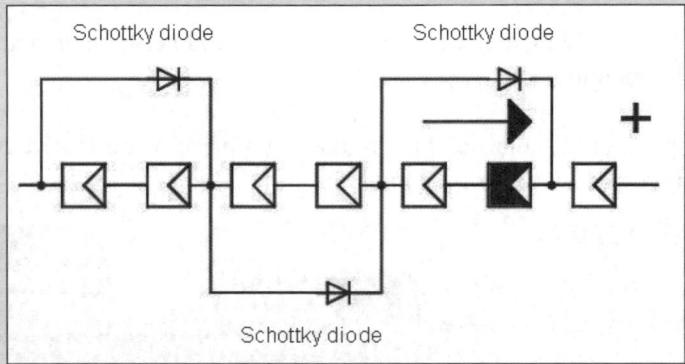

Figure 5.25: Bypassing the current via a Schottky diode when individual cells are shaded

5.6   Aligning the module with the light source

Required parts: solar module, breadboard, 10-Ω resistor (brown-black-black), 100-Ω resistor (brown-black-brown), orange LED, moving-coil gauge

Note

For the following experiments you need a bright light source (or full direct sunlight) for the solar module.

Figure 5.26: Aligning the setup with the light source

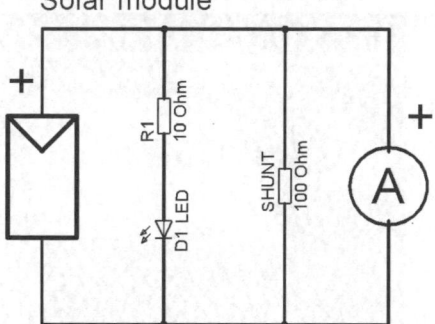

Solar module

Figure 5.27: Circuit diagram of the setup with orange LED and moving-coil gauge

Hold the solar module between your thumb and index finger (without shading the surface) and align the surface as perpendicularly to the light source as possible. What value does the moving-coil gauge show? Now move the module to vary the alignment and observe the gauge.

Note

The more perpendicularly the light beams hit the solar module, the more light energy can be converted into electric current to drive an electric consumer.

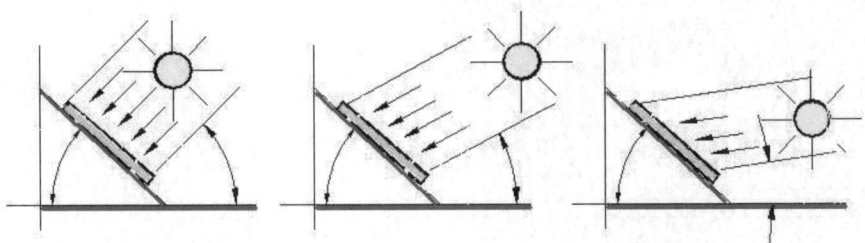

Figure 5.28: Angle of inclination between the surface of the solar module and the light beams. The number of arrows represents the intensity.

Lay some cardboard, wood or the like underneath the solar module to align it exactly with the sun. Observe the gauge. Wait at least an hour (or better several hours) and then have another look at the setup. The sun shines no longer perpendicular to the module, and the gauge shows a different value.

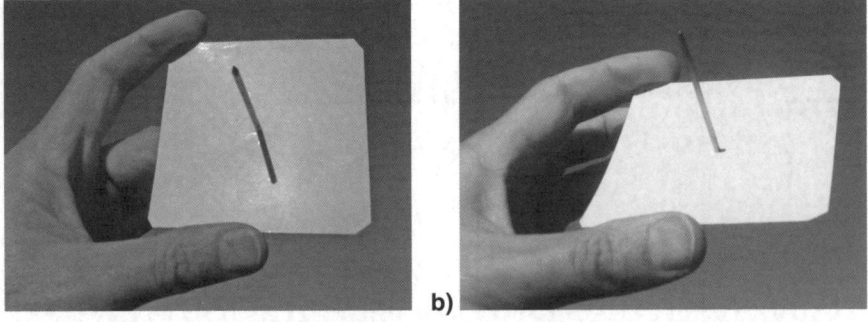

a)   b)

Figure 5.29: Perpendicular alignment aid: (a) a piece of cardboard with a match glued on (b) when the cardboard is aligned perpendicularly the match casts no shadow onto the cardboard.

Since the sun seems to move from the east to the west, you have to permanently move the solar module in order to track the optimal alignment with the sun.

5.7 Additional energy gain by using mirrors

Required parts: solar module, breadboard, moving-coil gauge, mirror (you can use reflective metal surfaces, mirror tiles, cosmetic mirrors, mirror film etc.). The mirror area must be at least twice as big as the surface of the solar module.

> **Note**
> For the following experiments you need a bright light source (or full direct sunlight) for the solar module.

The setup of the solar module and the measuring circuit is identical to that of the preceding experiments. When aligning the mirror you can see the mirrored light energy on the table, on the wall or on the module. The mirror must not shade the module. Please observe the current display and the LEDs when the mirrored light is cast on the modules.

a) Place the mirror in front and below the solar module

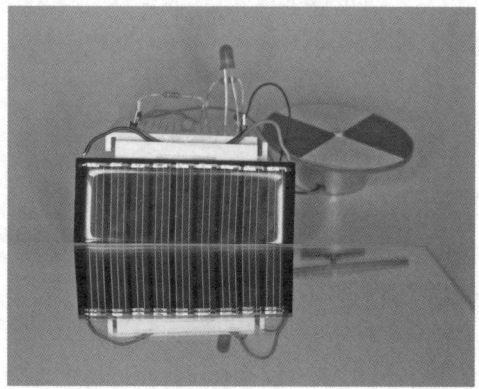

Figure 5.30: A mirror placed below the solar module

b) Place two mirrors on each side

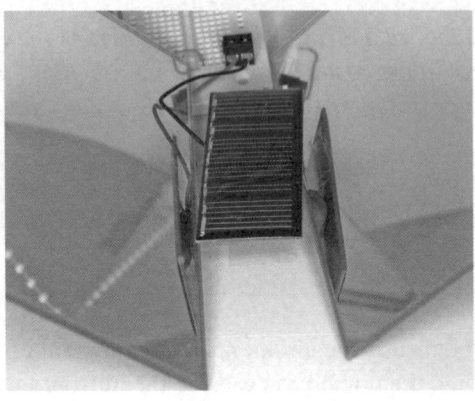

Figure 5.31: Mirrors on either side of the module

c) Use concave aluminium film or a concave mirror, e. g. a shaving mirror

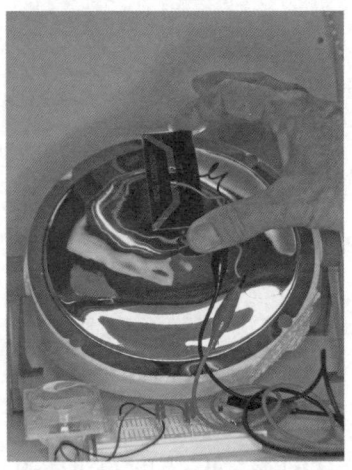

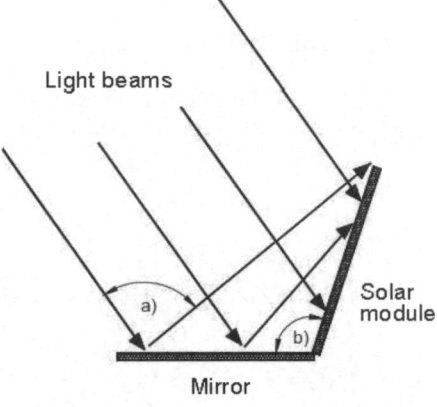

Light beams

Solar module

Mirror

Figure 5.32: The solar module in the centre of a parabolic mirror

Figure 5.33: Functional principle of using mirrors: The light beams reflected by the mirror towards the solar module carry additional energy. Note that the incidence angle has to equal the emergent angle.

When the mirror is aligned with the solar module at the proper angle, the light power transferred to the module is complemented by the mirrored portion. Thus you can increase the output current of a solar module in a simple way. However, this effect is limited by the fact that the mirror light warms up the solar cells which in turn reduces the performance. Because of this, it has been tried to use hybrid cells with mirrors or Fresnel lenses that divert the heat as thermal energy (z. B. for heating up water) and to take the electric current as electric energy with a good degree of efficiency.

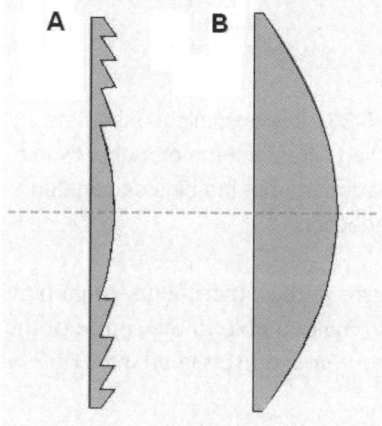

Figure 5.34: Structure of a Fresnel lens (a) compared to an ordinary lens (b). The Fresnel lens was originally invented by Augustin Fresnel to illuminate lighthouses. The shape of this lense offers the benefit that you need less material to focus the light beams so that you can produce the lens out of transparent plates.

5.8   Preparing the solar drive

Required parts: motor, breadboard, two pins, cardboard disc

The connection leads of the motor consist of flexible wire like those of the solar module. Connect the black and the red wire to the breadboard. It is recommended to insert the black end into the lower bar and the red end (+) into a contact of the five-contact row.

Figure 5.35: It is possible to insert the connection wires of the motor directly into the breadboard, but the pins can tighten the connection.

In order to recognise whether the motor shaft turns in the experiments you mount the cardboard disc on the shaft. Use a needle to make a hole in the centre of the disc and fit the disc on the shaft. Alternatively, you can use the adapter ring or the flap wheel included in the tutorial kit.

Figure 5.36: Preparing the cardboard disc for mounting it on the motor shaft

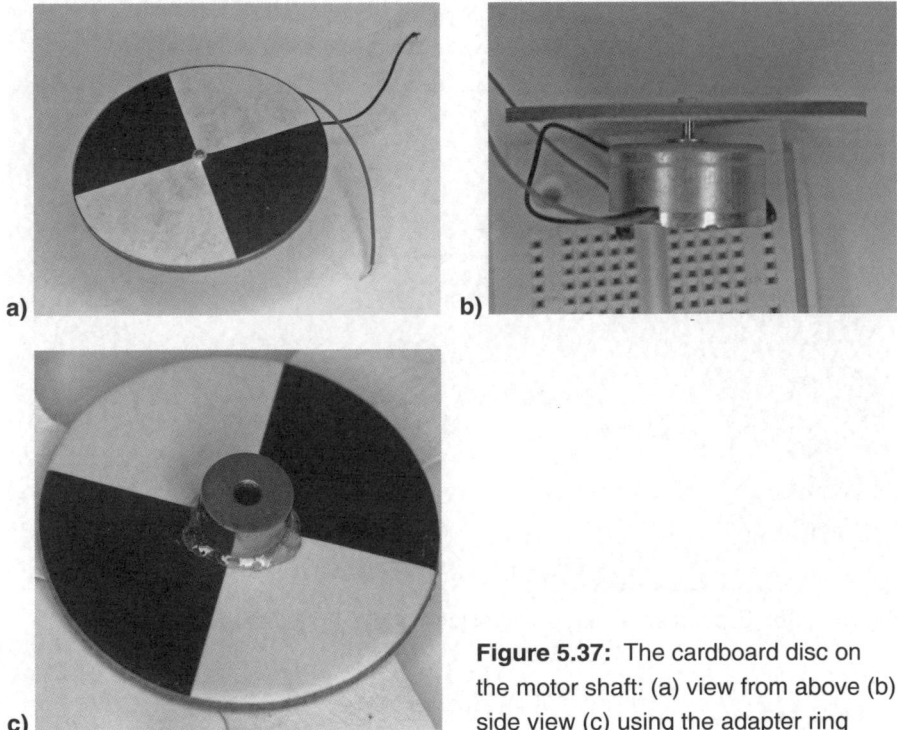

Figure 5.37: The cardboard disc on the motor shaft: (a) view from above (b) side view (c) using the adapter ring

5.9 Driving the motor by solar power

Required parts: solar module, breadboard, motor

> **Note**
> For the following experiments you need a bright light source or full direct sunlight for the solar module.

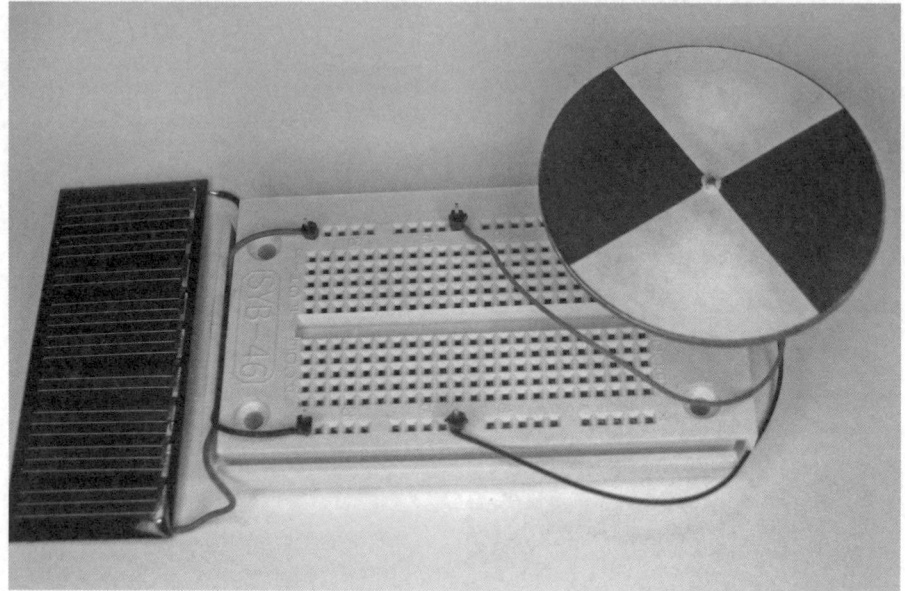

Figure 5.38: Experimental setup with solar module, breadboard and motor

Solar module

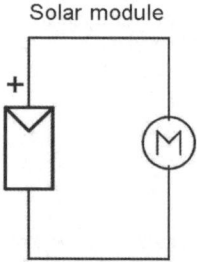

Figure 5.39: Circuit diagram
with solar module and motor

You can place the motor on the generator mount or you can attach it to a piece of cardboard using double-sided adhesive tape.

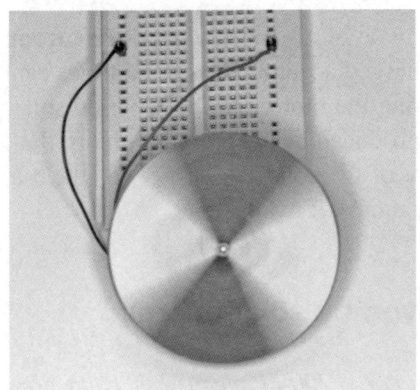

Figure 5.40: The disc rotates when enough sunlight is cast on the solar module

When enough sunlight is cast on the solar module, the motor shaft begins to turn. If the light is too weak, you will possibly have to crank the motor with your fingers so that it begins to turn. The reason for this is that the starting current of a motor can be more than twice as high as the operating current.

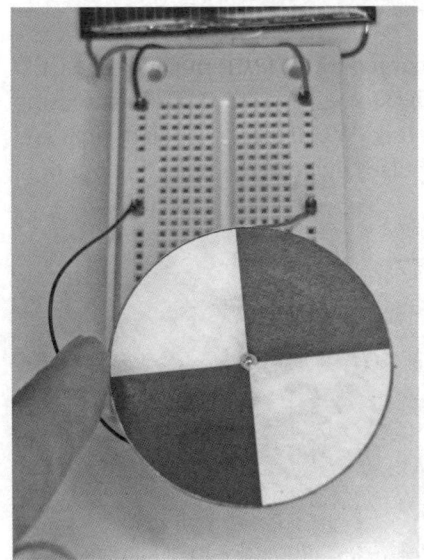

Figure 5.41: Cranking the motor with your fingers when there is not enough light. Reason: the starting current of a motor is higher than the operating current.

This experiment shows how the operating mode of a solar module differs from that of an accumulator or a battery. Fully charged accumulators or batteries can easily provide the required current for starting the motor. A directly used solar module can only provide the current that is produced by the present light irradiation (depending on the degree of efficiency of the solar cells). Try to attach a 1.5-V battery or accumulator directly to the motor, if availabe.

5.10   Jump-starting a motor driven by solar energy

Required parts: solar module, breadboard, motor, 4,700-μF electrolytic capacitor, flashing LED

> **Note**
>
> For the following experiments you need a bright light source (or full direct sunlight) for the solar module.

In this circuit, the electrolytic capacitor is charged by the solar module. The LED and the motor are connected in series with the storage capacitor. With increasing charge of the capacitor, the LED begins to blink. When there is enough light and energy flow, the motor receives electric impulses that cause a pulsating rotation.

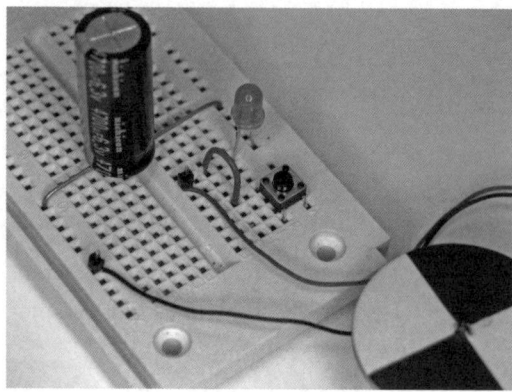

Figure 5.42: Setup on the breadboard with electrolytic capacitor, LED and push-button

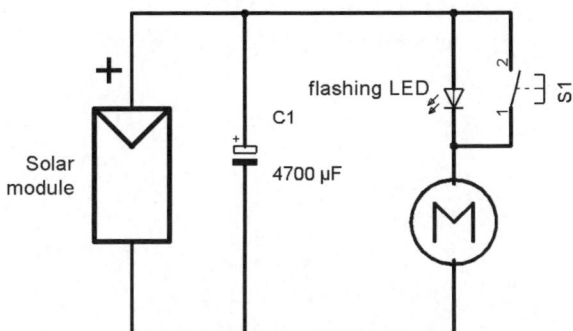

Figure 5.43: Diagram
of the jump-start circuit

You can now connect the motor with the electrolytic capacitor via the push-button. When the capacitor is charged, the disc turns at high speed.

Additional experiments: Experiment with the setup and use it with and without the push-button and with one additional resistor. Try out various resistors with resistance values of 10 Ω (brown-black-black), 100 Ω (brown-black-brown) and 1 kΩ (brown-black-red). How does the motor speed and the mode of operation change?

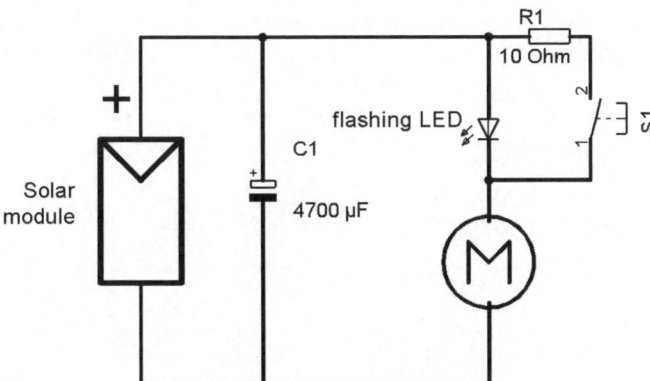

Figure 5.44: Additional experiment with R1 resistor

These additional experiments according to the setup of Fig. 5.44 show that the current flow to the motor is changed by the resistors. In this way you can also control the speed.

6 Wind Power

Besides solar energy, wind power is another important renewable energy source. Wind energy has played an important role in the history of mankind and it has become more and more important to provide renewable energy. In the past, grain was grinded in windmills, and even today many farmers all over the world still use water pumps driven by wind power.

6.1 What is wind power and how is it generated?

Wind is caused by solar radiation and three other basic factors (as seen from a global perspective):

- The solar radiation hitting the earth is stronger at the equator (perpendicular incidence angle) than in the polar regions (flat incidence angle). Thus the air near the equator is heated up more intensely and rises, while cold air flows from the colder regions (temperate zones) to the equator (*Hadley circulation*).

- Accordingly, there is a flow of air between the poles and the temperate zones, called the *Rossby circulation*.

- The third factor consists of the currents that are caused by the heating and cooling of huge masses of water and earth.

The direction of the wind currents is affected by the rotation of the earth. Regional winds, too, are caused by different temperature levels and other factors like different storage capacity, building etc.

When wind power is used as a source of energy, two questions (referring to the site) are of particular interest: How strong and how often does the wind blow?

In order to answer these questions and thus determine whether a location is suited for generating wind energy, you can request a wind zone map at the met office (www.metoffice.gov.uk). These maps indicate the annual average of wind speeds (for 10 m, 25 m and 45 m above ground level) and thus give you a rough lead where you can expect how much wind in the course of the year.

Suitable sites for erecting a wind engine are among others hilltops and draughty valley bottoms. Because of possible turbulences, a location on a slope is problematic for blade rotor models but suited for Savonius rotors. As wind generators are more or less noisy depending on the model you should also consider the neighbourhood.

Of course there are some questions you have to answer up front: What minimum wind speed does the wind engine require to produce electric current? What machines shall the generator drive? How hard does the wind blow in average? Are there many stormy days so that it is worth to build a storm wind generator, or is this a location a weak wind zone? In any case, it is reasonable to monitor and measure the wind regime over a longer period of time before you invest much effort to erect a complete wind generating plant.

Not only when sailing but also when using renewable wind power, it is important to agree on the meaning of wind forces. Wind speeds are given in various units depending on the country: you can see specifications in knots, metres per second, kilometres per hour, miles per hour and Beaufort. All units except those of the Beaufort scale can be converted into each other.

Originally, the Beaufort scale had no direct reference to the wind speed but consisted of 13 steps describing the effect of wind on the big sailing ships of those days. Not before the middle of the 19th century an official correlation between the Beaufort values and the wind speeds was defined. The wind speeds given in this definition are given for 10 m above ground level.

Beaufort value	Visible signs	Wind speed in m/s	Wind speed in km/h
0	Calm, smoke rises vertically	0–0.5	0–1.8
1	A light air can be felt	0.6–1.7	2–6
2	Smoke slightly drifts away	1.8–3.3	6.5–12
3	Leaves lightly move	3.4–5.2	12.5–19
4	Small twigs move	5.3–7.4	19–27
5	Small branches move	7.5–9.8	27–35
6	Large branches move; audible, strong wind	9.9–12.4	35.5–44.6
7	Small trees move, wind howls	12.5–15.2	45–55
8	Big trees move	15.3–18.2	55–65.5
9	Storm, roof tiles are thrown down	18.3–21.5	66–77.5
10	Trees are uprooted	21.6–25.1	78–90
11	Severe damages	25.2–30	90.5–105
12	Hurricane, most severe damages	more than 33	119

Modern anemometers usually give the wind speed in metres per second. In the news you often hear gale warnings with specifications in kilometres per second. The equivalent values are given in the third and fourth column of the table. In order to convert the values yourself, you use the formula: (value in m/s) x 3600 / 1000 = (value in km/h).

6.2 Mode of operation of the electrical wind generator

The following sections will make you acquainted with the properties and the operation of a wind generator by means of practical experiments. You will learn how to exploit wind and what you have to keep in mind in order to gain an optimal energy yield. In the following text the electrical machine is called a *generator* when it is used to convert the mechanical energy provided by the wind into electrical energy. Contrarily, when it is used to convert electric current into mechanical energy, it is called a *motor.*

Connecting the generator to the breadboard

Required parts: generator, breadboard, two pins

The electrical machine has two connectors with a red and a black wire, respectively. The generator provides direct current and therefore has a plus and a minus pole, just like a battery. Connect the black and the red wire to the breadboard. It is recommended to insert the black wire into the lower bar und the red one into the upper bar (see Fig. 6.1a).

For most of the following experiments you can leave the generator connection unchanged.

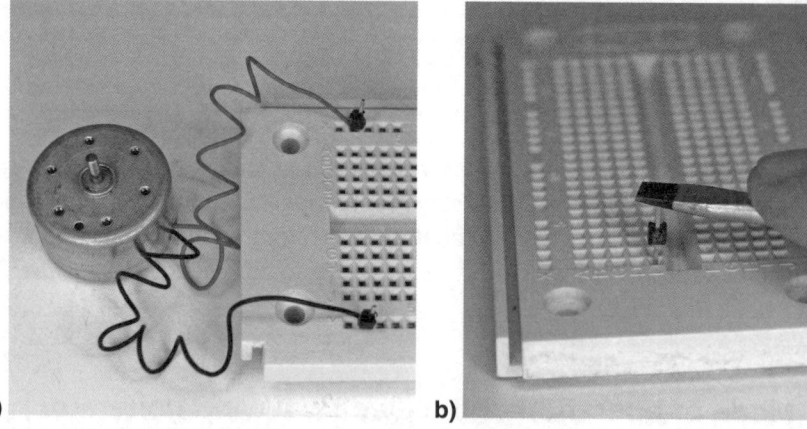

a)   b)

Figure 6.1: (a) Stabilising the connection wires of the generator with pins (b) pressing down the pins with the help of a screwdriver

6.3 First experiments with the generator

Required parts: generator, breadboard, red LED, orange LED, wire jumpers, adapter ring (brass).

In addition to the generator, you now connect the orange LED to the breadboard. The longer leg is the plus pole and has to point towards the red generator connector. Furthermore, you need a wire jumper in order to connect the LED with the lower contact row. When all components are inserted, you can try to turn the generator shaft with your thumb and your index finger. When you turn the shaft in one direction, the LED will flash up. In the other direction nothing happens.

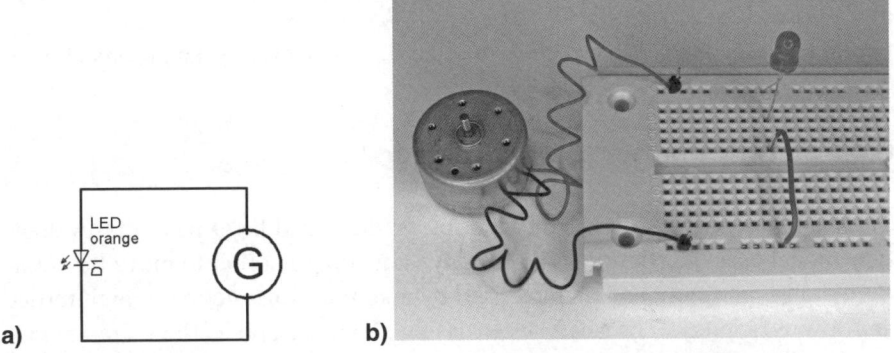

a)   b)

Figure 6.2: Generator with LED (a) circuit diagram (b) setup on the breadboard

In the next step, you insert the red LED, this time with the longer leg pointing towards the black generator wire. Again you have to connect the LED with the lower contact row using a wire jumper. When you now turn the generator shaft with your thumb and your index finger, the red LED flashes in the one direction and the orange LED in the other.

Optionally you can slip the adapter ring over the generator shaft and secure it with the screw. Now you can turn the shaft using the ring and observe the LEDs.

The experiment shows that the mechanical movement is converted into electrical energy. Depending on the direction of the rotation, the polarity changes. This is indicated by the two LEDs that are connected to the breadboard with opposite polarity.

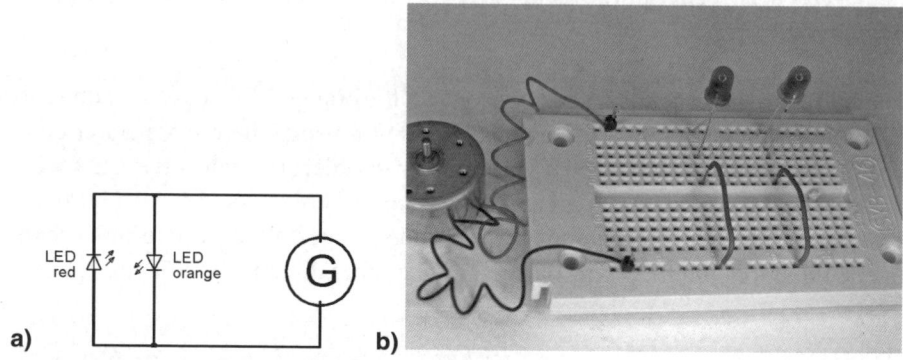

Figure 6.3: Generator with two LEDs (a) circuit diagram (b) setup on the breadboard

6.4 Requirements for a wind rotor

There are basically two types of wind rotors: drag- and lift-type rotors. A drag-type model uses the friction of the aerodynamic drag in order to move the rotor. Pure drag-type rotors are characterised by a simple construction, a high torque and low efficiency. The shaft speed is low. Typical applications are anemometers.

Lift-type rotors use the lift effect as airplane wings do. They are delicate objects that sometimes look futuristic. This type of wind power plant offers the benefit of exploiting the wind energy in a much better way than drag-type rotors do, so that they yield a higher output.

Aligning the axis

There are basically two configurations of wind power plants: those with a vertical and those with a horizontal axis. Historically, the vertical-axis models are the older ones.

The horizontal-axis variants comprise rotors with one, two and three blades, Dutch windmills and the classical American wind turbines.

The single bladed rotor has only one blade that is balanced by a counterweight. The balance of the dual and triple bladed rotors is achieved by the alignment of the blades. The most popular vertical-axis wind energy converter is the Darrieus rotor.

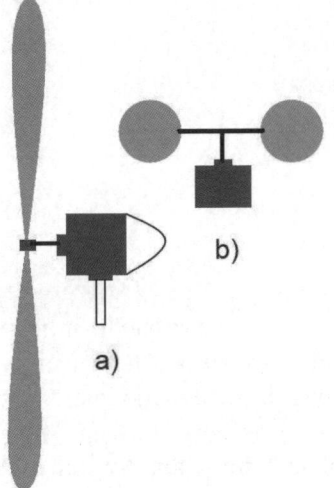

Figure 6.4: Types of axis alignments
(a) horizontal axis (b) vertical axis

On the first sight, the rotor of a wind generator looks like the propeller of an aircraft, but there is a basic difference between these two constructions. An aircraft propeller presses the air backwards in order to achieve propulsion while the wind rotor operates in the opposite direction. The profiles of the blades are twisted and built the other way round. It may seem a good idea to drive a wind generator by an aircraft propeller but it will not work, even if you install the propeller in the reverse way. The simple-profile blades of fans can be utilised as wind rotors, but their use is very limited. They offer only a low degree of efficiency and are only suited for drag-type rotors.

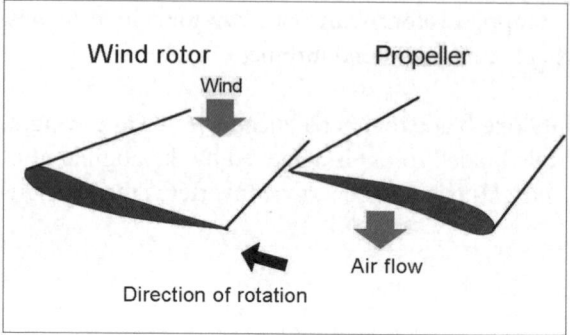

Figure 6.5: Basic principle of a wind rotor vs. a propeller

6.5 Building the generator housing

Required parts: generator bracket, glue

The generator is installed in a generator housing so that you can handle it more easily in the wind power experiments. First you have to cut out the generator bracket in the included cardboard template. Cut along the outer solid line.

Mark out the dot-dash lines (folding lines) with a blunt knife so that you can fold the cardboard easily. Fold the generator housing into a triangular body and glue strap B under edge A. Glue area C to the lid of a water bottle. Let the construction dry completely. A more durable connection between the housing and the lid can be achieved by using double-sided adhesive tape.

Note

The best practice when using liquid glue is to initially apply glue only to one of the two surfaces that have to be connected. Place the glued area onto the other one, move it a little to and fro and separate the two areas immediately. Let the glue on both surfaces dry a little and then press them together.

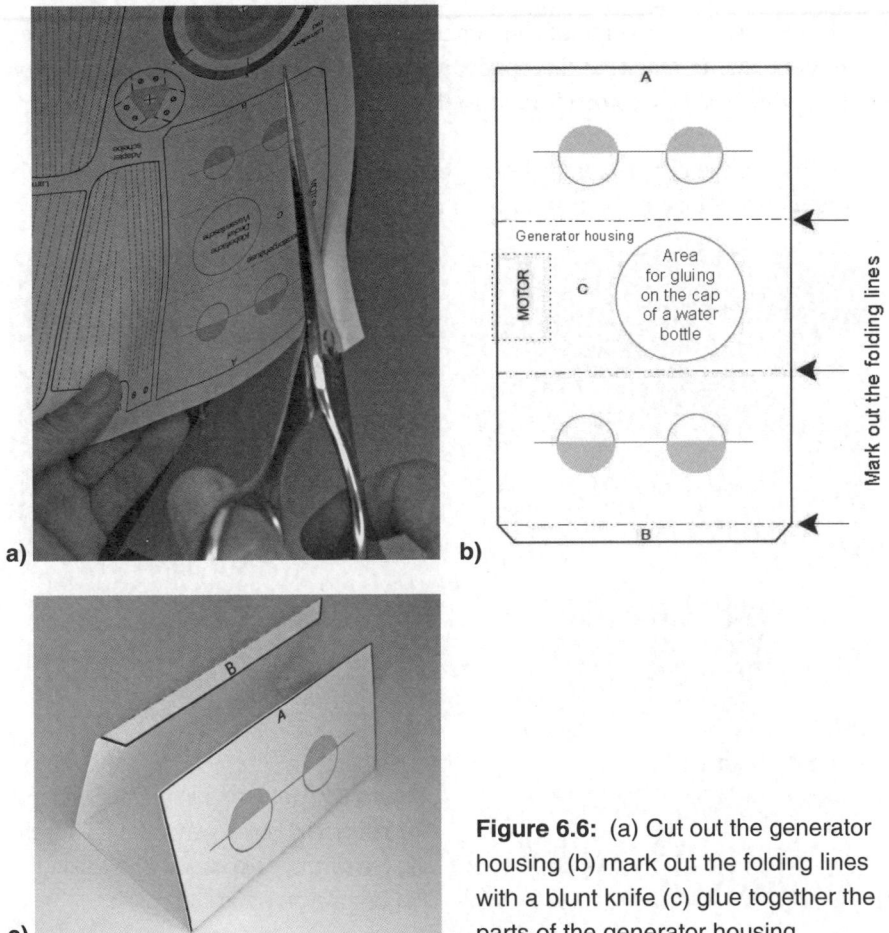

Figure 6.6: (a) Cut out the generator housing (b) mark out the folding lines with a blunt knife (c) glue together the parts of the generator housing

Next you glue the generator into the triangular housing so that approx. 2–3 mm of the metal casing stick out of the cardboard housing and the connection wires can be placed freely in the triangular cavity. Make sure that no glue ends up in the motor/generator or jams the axis. As most glues for handicraft work do not stick well onto the metal surface of the generator, you can use double-sided adhesive tape in order to achieve a more durable connection between the generator and the housing.

Attach the connection wires of the generator with the alligator-clip cables (red to red and black to black) at the opposite end of the housing. As the connection wires are thin it is best to bend them so that the alligator clips have a good grip.

Depending on the experimental setup you can also use the alligator clips to fix the wire connections of the generator to the housing.

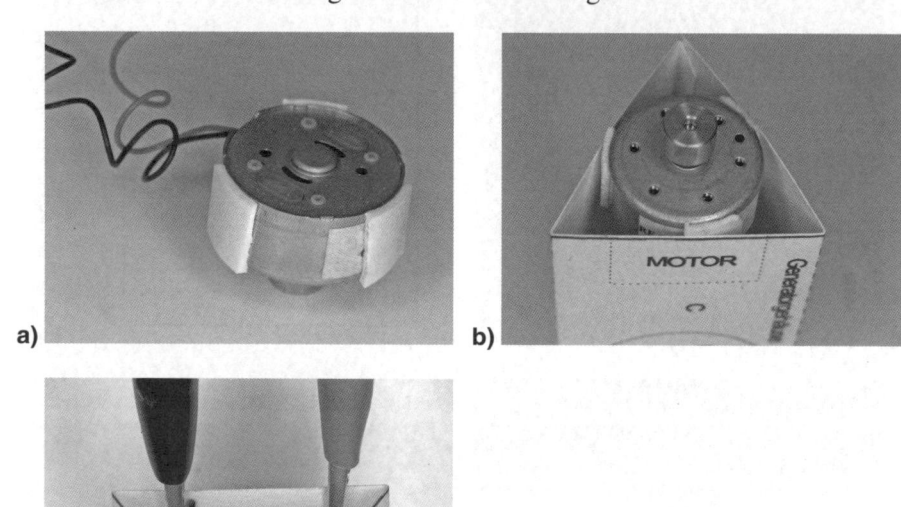

a)

b)

c)

Figure 6.7: (a) Attaching adhesive tape (b) installation of the generator (c) attaching the alligator-clip cables and locking the connection wires into position

6.6 The base (pylon) of the wind rotor

Required parts: generator in housing, water bottle with plastic lid (0.35–0.5 l), alligator-clip cables (red, black), glue (and/or double-sided adhesive tape)

The pylon of very big wind rotors is also called a 'tower'. The higher the rotor is situated above ground level, the more wind energy can be utilised. A 10 % increase in wind speed yields 33 % more power. (The power increases with the third power of the wind speed.) Of course, the tower has to be robust enough so that the wind cannot blow it down.

In the experiments you can use a small plastic water bottle with plastic lid as the pylon. Fill the bottle up to the half with water so that it has a low centre of gravity and the wind rotor stands firmly on the ground. Big 1.5-l bottles are not suited because the rotors are impeded in their movement by the bulky body of the bottle.

Note

Make sure that the water bottle stands firmly by filling it halfway with water before you put on the lid (with the generator housing glued on).

The lid can be removed from the bottle at any time, so you can use the housing and the generator without the pylon.

In strong wind you have to secure the bottle by additional means. Otherwise it may be blown down which would damage the rotor.

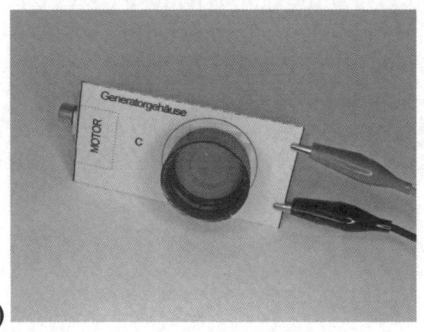

a)   b)

Figure 6.8: (a) Lid glued to the generator housing (b) plastic bottle with generator housing

6.7   Building the drag-type rotor

Required parts: cardboard disc (flap wheel), six cut-out flaps, adapter ring (brass)

In order to build the drag-type rotor you have to cut out the disc and the six flaps in the included cardboard template. Cut along the outer solid lines. The radial lines marked with 'x' represent slots you have to cut in with scissors. Attach the flaps to the disc by inserting the slots into each other. The bevelled corner points to the rear. After sticking the parts together you can stabilise the connections additionally with some glue. Before the glue is completely dry you have to adjust the flaps again. Now attach the brass adapter ring to the cardboard disc either with glue or with a piece of double-sided adhesive tape. To place the ring exactly in the middle of the wheel it is useful to pierce a needle through the cross line designation on the front side of the disc (the centre) and to attach the adapter ring (with some sort of adhesive) in the centre using the needle as a guide. Let the glue dry.

Note

In subsequent experiments the drag-type rotor is used as a water wheel.

Note

The wind rotor used here works according to the aerodynamical principle of draft. It can only be used in natural wind when one side of the wheel is shielded against the wind or when the wind is directed by a 'funnel' to one side. In the following experiments with a targeted air stream, it is used as an illustrative model.

a)

b)

c)

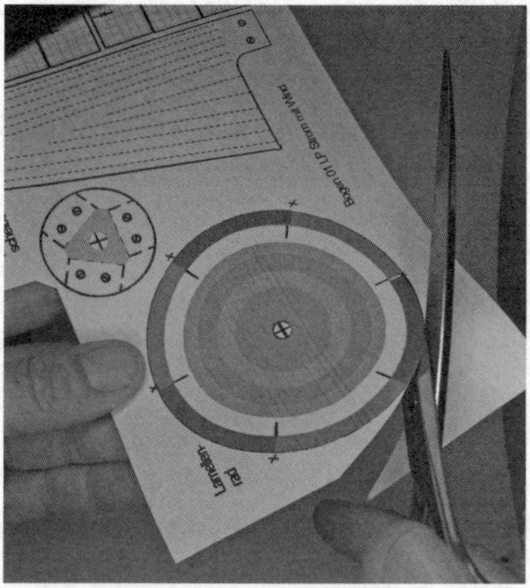

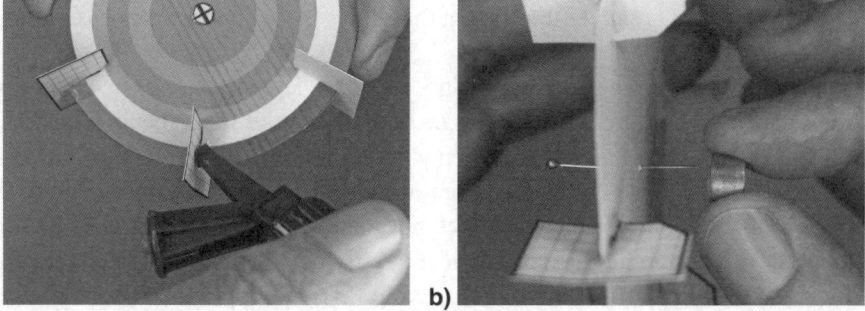

Figure 6.9: Drag-type rotor: (a) cutting out (b) cutting in the slots (c) attaching the flaps with the disc

a)

b)

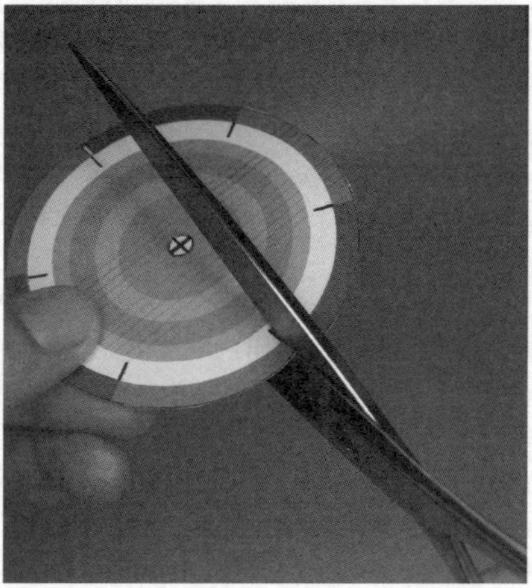

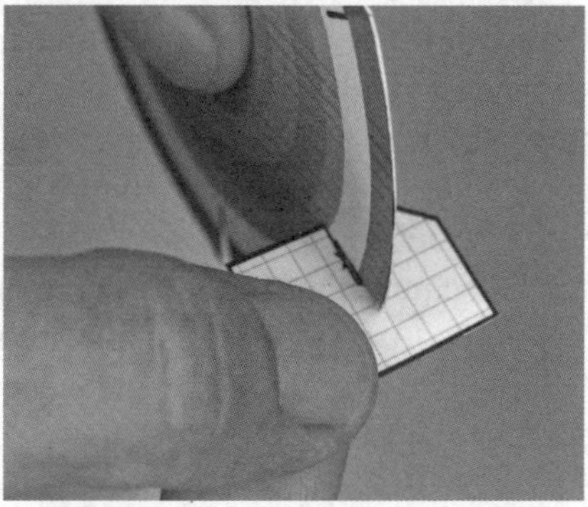

Figure 6.10: (a) Adding some glue to the connections (b) gluing on the adapter ring

6.8   Making wind with the drag-type rotor

Required parts: generator/motor, breadboard, two pins, flap wheel, adapter ring, battery clip, 9-V battery, orange LED, 470-Ω series resistor (yellow-violet-brown), 100-Ω resistor (brown-black-brown), push-button

For a start, we try to produce wind. Attach the flap wheel that you assembled in the preceding step to the motor shaft so that it can turn freely.

Then insert the connection wires of the battery clip into the breadboard. You may secure them additionally with a piece of adhesive tape. Insert the 470-Ω series resistor, the orange LED, the 100-Ω resistor, the push-buttons and two pins for the connection of the motor by the alligator-clip cables into the breadboard.

Now you can connect the battery to the battery clip. The LED lights up first and the wheel jerks a little or begins to turn (depending on the start-up behaviour of the motor). If the wheel does not turn, press the push-button for a short while so that the motor receives an electric impulse and the wheel begins to turn fast. After releasing the button, the rotation of the wheel becomes slower. You can feel the difference in the produced 'wind'. By disconnecting and reattaching one of the alligator clips you can repeat the process. This time you can jump-start the wheel with your fingers. The motor will continue to turn driven by the electric current that is reduced by the amount needed for the LED.

In this experiment you can see that the motor needs a higher start-up current. After it is started it can be driven by the smaller electric current that flows through the LED. The LED reduces the current that flows to the motor to approx. 14 mA.

You can experience the generated 'wind' directly: depending on the current flow to the motor you can feel (a) a light air flow above the LED rectangular to the flap wheel (b) a strong air stream when you press the push-button. You can swap the motor connections (alligator clips) at the pins. When producing wind, the polarity of the motor connection does not matter. However, it is possible that the motor runs smoother or starts up more easily when you swap the connection. The reason for that is that the motor used in this kit is optimised for one direction of rotation.

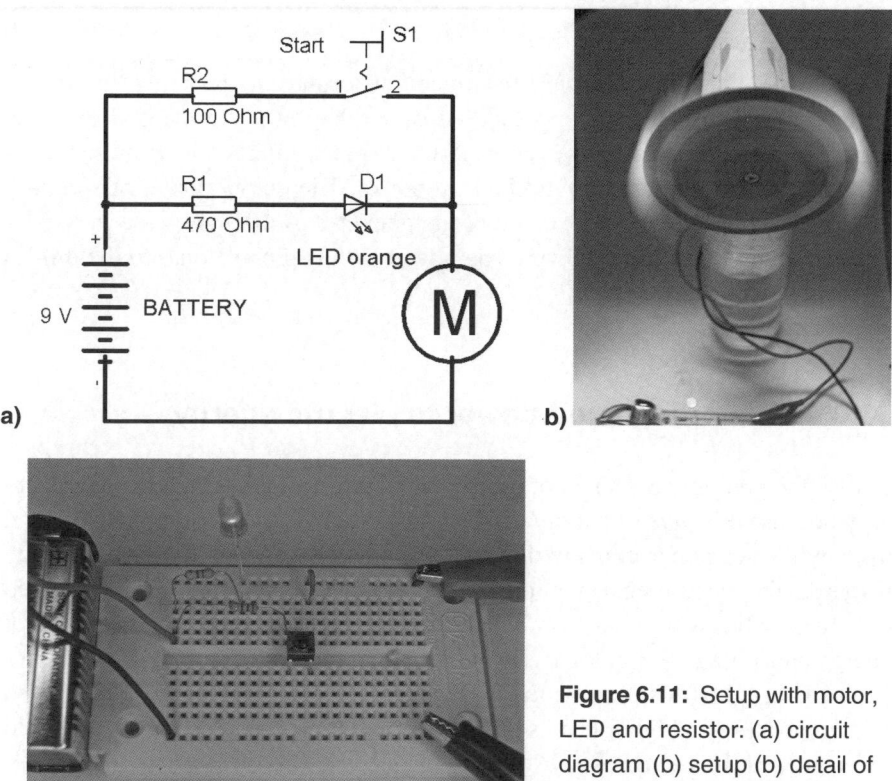

Figure 6.11: Setup with motor, LED and resistor: (a) circuit diagram (b) setup (b) detail of the circuit

In Fig. 6.11c you can see that only the two upper connections of the push-button are used. One is attached to the 100-Ω resistor, the other one to the wire jumper.

6.9 Using wind power

In principle nearly all wind rotors can be used for a direct mechanical drive as well. This is a well-known mode of operation that is used in windmills to grind grain or to saw wood. It is also employed in the traditional Greek oil mills with covered blades and in the traditional American wind turbines and Savonius rotors used for pumping water or for ventilating a pond.

Performance of wind rotors

The bigger the contact surface, the higher the torque and the better the start-up performance in weak wind conditions, but the lower the speed. The bigger the wheel, the lower the speed. A wind rotor cannot convert more than 2/3 of the wind power into mechanical power. The output power of a wind rotor is increased with the square of the diameter. A 10 % increase in wind speed yields 33 % more power. The power is thus proportional to the third power of the wind speed.

6.10   Converting wind power to electric energy

Today, the predominant share of utilised wind power is converted to electric energy because this type of energy can be transported via the existing network and used by the consumer in many different ways. Because wind is a very unsteady resource, the electric energy gained by wind power plants can only be utilised in a reasonable way when combined with the national grid, other sources of energy and storage technologies. In isolated operation the use of storage devices is mandatory. For this purpose the wind-generated electrical energy can be used to pump water to a higher level so that the stored kinetic energy can be retrieved when needed. It is also possible to store the electric energy in accumulators.

Gear transmission

Usually a transmission is used in big wind power plants to adjust the low speed of the rotor to the required high speed of the generator. The drawback is that the transmission consumes mechanical energy and impedes the start-up of the rotor.

Anyone who tried to turn a motor with attached transmission knows that this is harder than without a transmission. The higher the transmission ratio, the more torque is needed to turn the drive shaft.

In our experiments no transmission is used. The generator is directly attached to the rotor.

Preparing and using the gauge

Required parts: Breadboard, moving-coil gauge

To perform the subsequent experiments, insert the connection leads of the gauge into the contacts of the breadboard – the black wire (minus pole) into the lower bar and the red wire (plus pole) into one of the upper contacts. The ends of the connection leads are tin-plated and can directly be inserted into the breadboard contacts.

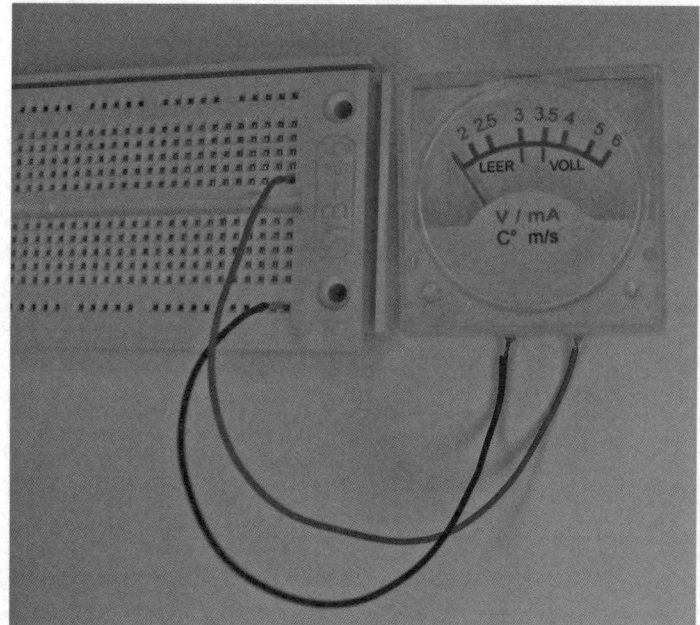

Figure 6.12: Connecting the gauge to the breadboard

Note
Never connect the moving-coil gauge directly to a battery because this could damage it.

6.11   Experiments: Measuring the voltage of wind-generated electricity

Required parts: generator with housing and flap wheel, breadboard, alligator-clip cables (one red, one black), pins, moving-coil gauge

> **Note**
>
> Each experiment in this section builds upon its predecessor. Thus, it is not necessary to dismount the prior setup. Instead, you add, remove or replace individual parts to extend it.

Insert one pin into the lower and one into the upper bar of the breadboard. Attach the generator via the alligator-clip cables to the pins, for the time being without considering the polarity (minus/plus). Puff at the flap wheel from above, from the right side and from the left side so that the direction of rotation changes. Monitor the gauge.

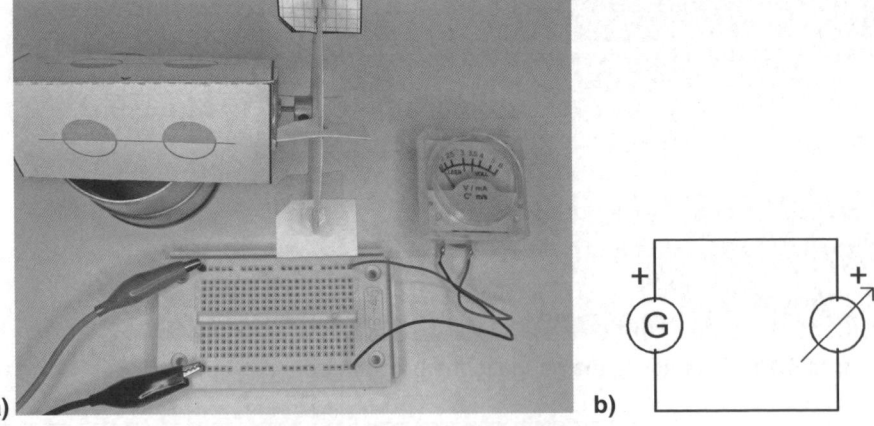

a)   b)

Figure 6.13: (a) Setup with flap wheel (b) circuit diagram

Depending on the polarity of the generator connection to the gauge and the direction of rotation, the needle of the gauge is deflected to the left or the right. When it is deflected to the left, the needle is blocked mechanically by the stop.

In this direction you cannot read a value because of the missing scale. Therefore you now extend the circuit by a diode so that the moving-coil gauge only receives electric current in the one direction that causes a usable needle deflection.

Required parts: as before, plus an additional diode (glass body with black ring) inserted according to the picture. For now, the mounting direction does not matter.

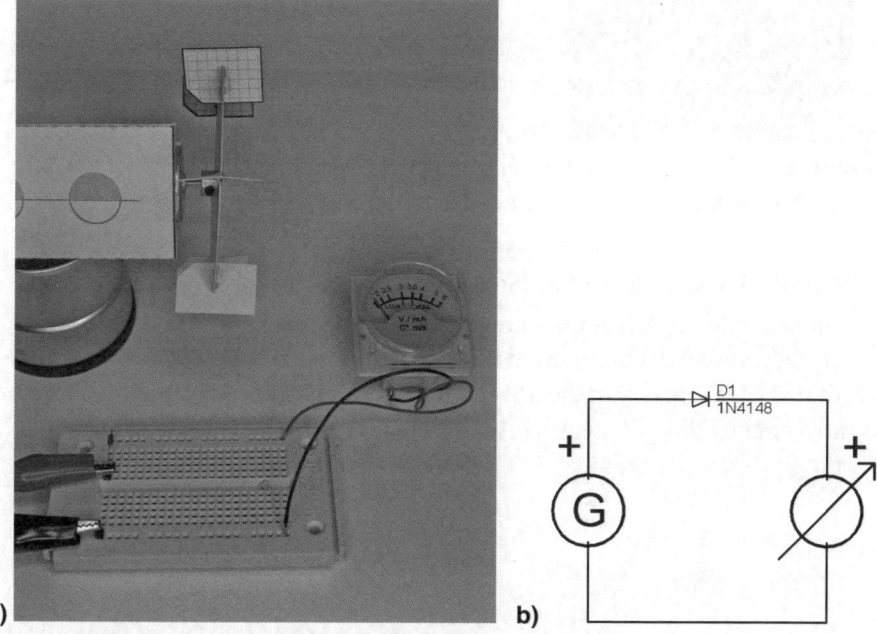

a)   b)

Figure 6.14: (a) Setup with silicon diode (b) circuit diagram

Now the needle of the moving-coil gauge is always deflected to the same side irrespective of the direction of rotation of the flap wheel. The direction of deflection depends on the polarity of the generator connection and the diode.

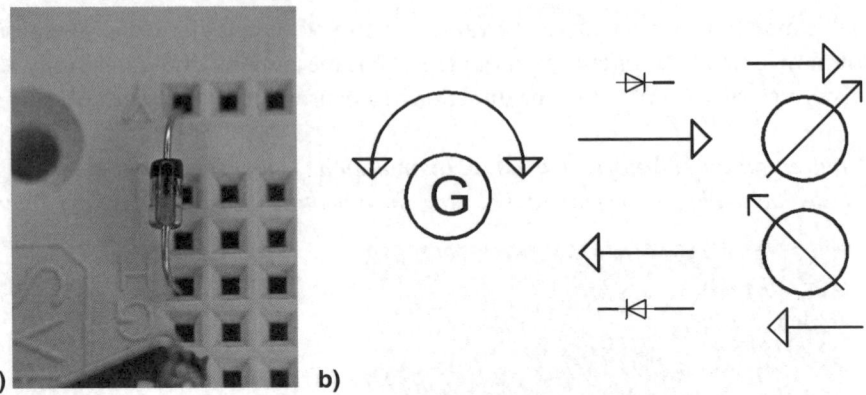

a)   b)

Figure 6.15: (a) Mounting direction of the diode (b) basic possibilities for the direction of the electric current and the needle deflection

---

Technical and actual direction of electric current

The electrons flow from the minus pole to the plus pole. This is the *physical direction* of electric current as opposed to the *technical direction* that is defined as leading from the plus pole to the minus pole. In circuit diagrams the technical direction is usually depicted (e. g. for LEDs, diodes and generators).

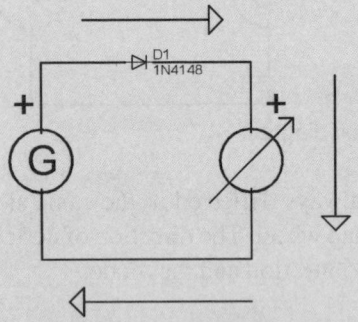

Figure 6.16: Conceptual representation of the technical direction of electric current (in a closed circuit)

The diode works like a valve that only allows an energy flow in one direction. In the other direction, it blocks.

Connect the generator via the alligator-clip cables to the breadboard with the correct polarity and insert the diode so that the bar on the casing (cathode) points towards the moving-coil gauge. When you now blow hard at the flap wheel in one direction, the needle of the moving-coil gauge is deflected up to the right stop. Turning the wheel in the other direction causes no needle deflection at all.

Furthermore, you can feel that the wheel is harder to turn in the direction that causes a deflection, because the resistance of the generator increases due to the electric power consumption of the gauge.

In the next step you will modify the measuring instrument so that you can read off the values easily.

Required parts: as before plus an additional 4700-µF electrolytic capacitor

Experiment by connecting successively a 10-kΩ resistor (brown-black-orange) and a 1-kΩ resistor (brown-black-red) in parallel with the capacitor.

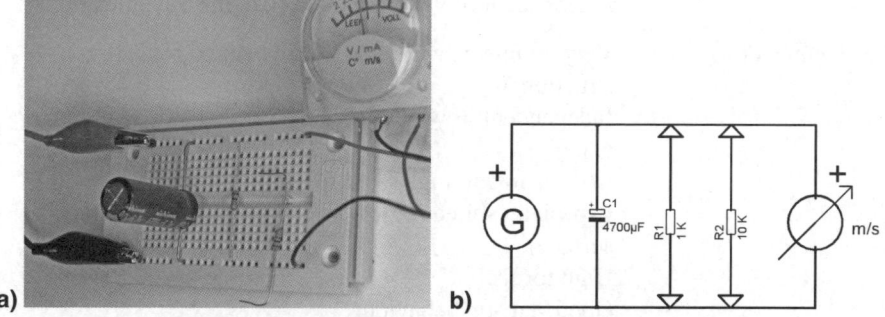

Figure 6.17: (a) Setup with capacitor and resistors (b) circuit diagram

The needle is now deflected more smoothly and you have more time to read off the displayed value. The wheel charges the capacitor, and the resistors discharge it. The discharge time of the capacitor depends on the resistance value. This influences slightly how fast the needle moves back to '0'.

6.12   Different types of wind rotors

In the course of history, many different types of wind rotors have been developed. The following table provides an outline of the most common types and their properties:

Wind rotor type	Benefits
Traditional American wind turbine	Simple blade profile Good start-up behaviour High torque Suited for directly driving a water pump
Blade rotors	High speed High degree of efficiency according to the number of blades Suited for directly driving a generator, delicate structure, silent running
Darrieus rotors	Delicate structure, good degree of efficiency Independent of wind direction (like the Savonius rotor)
Savonius rotors	Very simple construction, ideally suited for self-construction Independent of wind direction (vertical axis without vane) Utilises low to high wind speeds Especially suited for low wind speeds (huge contact surface) High torque Good start-up behaviour No problems in gusty conditions

6.13 Building the blade rotor

In the following sections you will build a triple bladed wind rotor and experiment with it.

> **Note**
> Each experiment in this section builds upon its predecessor. Thus, it is not necessary to dismount the prior setup. Instead, you add, remove or replace individual parts to extend it.

Required parts: Wind rotor blades and adapter disc on the cut-off cardboard, cardboard disc, adapter ring (brass), glue

Cut out the three rotor blades on the included cardboard template. Cut along the outer solid line. In order to profile the blades, draw them face up across the edge of the table or the like so that they get a round shape. You can also roll them between your thumb and index finger until they show the desired profile. It is no problem when the curvature is at first too extreme because you can reduce it when experimenting. However, there must not be any sharp bends. Glue the inner ends of the blades to the adapter disc. For this purpose, apply glue to the area on the disc marked by the dotted lines, put the blade end onto the glue, push it a little to and fro and remove it. Let the glue dry on both parts and press them together.

After having performed all the experiments with the flap wheel, you can now carefully remove the brass adapter ring. For the following experiments you have to glue the ring to the centre of the blue-white cardboard disc. To this end you can attach a piece of double-sided adhesive tape to the adapter, stick a nail through the ring and target the cross lines on the disc with its tip. It is also possible to apply a contact adhesive to both parts, let it dry for some minutes and press the parts together. When attaching the disc to the motor shaft, the blue-white surface should point 'backwards', i.e. towards the motor casing. The adapter disc with the three rotor blades is glued to the grey side. Follow the gluing guidelines as described above (with interim drying).

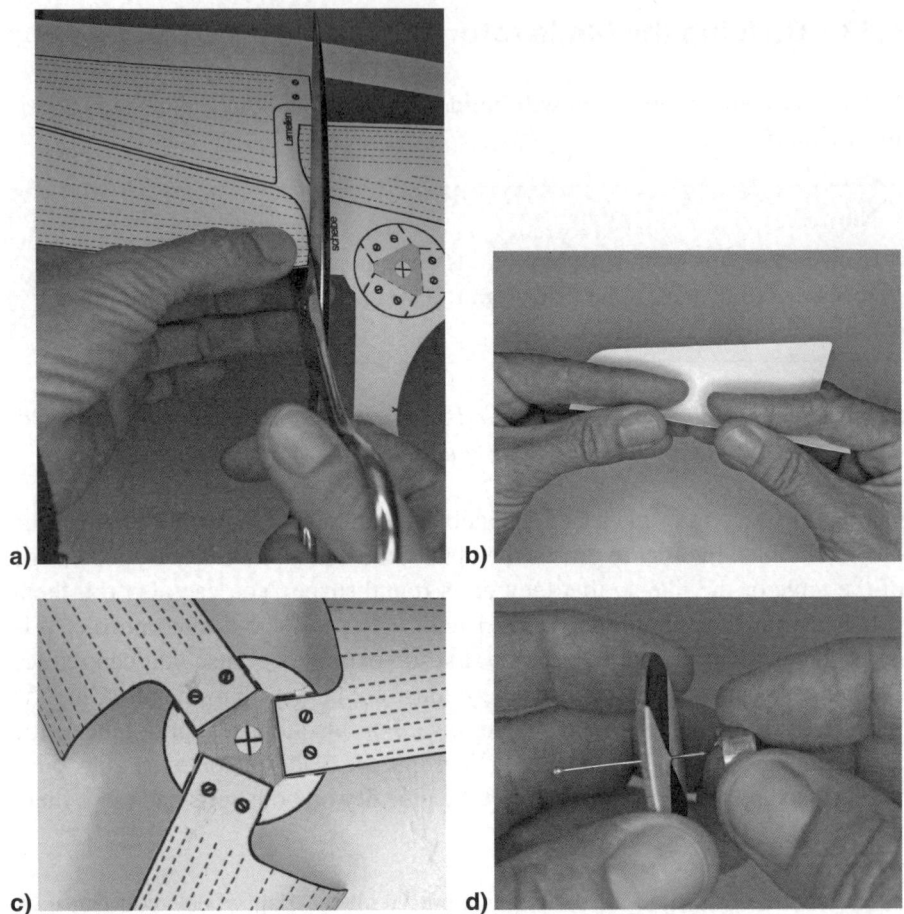

Figure 6.18: Assembling the wind rotor from the individual blades: (a) cutting out the parts (b) shaping the blades (c) gluing the blades onto the adapter disc (d) gluing the brass adapter ring to the cardboard disc

Figure 6.19: Finished rotor with cardboard disc and adapter ring as seen from behind

The profile of the rotor blades

Why is it important to shape the profile of the rotor blades? The blades of a lift-type rotor have a shape similar to an aircraft wing. When air flows towards the front edge of an efficient rotor blade profile, it moves faster along the back side because the way is longer due to the curvature. The way along the front side is shorter, therefore there is a higher pressure than on the back side. This pressure difference between front and back side causes lift. At the same time the rotor blade experiences an aerodynamic drag. The ratio between lift and drag is called *glide ratio* (or simply *lift-to-drag ratio*). The lift should be higher than the drag.

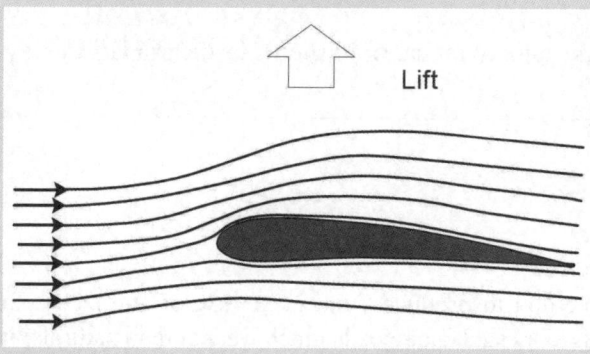

Figure 6.20: Air flow around the profile of a rotor blade profile

To drive the finished rotor, air is directed to the unprinted side of the blades.

6.14 Experimenting with the rotor

Required parts: generator, generator housing with wind rotor and adapter ring, installed on a bottle

Now you can attach the whole wind rotor to the generator shaft. The generator housing sits on the bottle. Take the bottle in your hand and blow at the wheel from different directions (from the front, from behind, sideways), at first without connecting the generator to the breadboard. What do you observe? When do the blades exhibit a more distinct movement? When do they rotate not so easily?

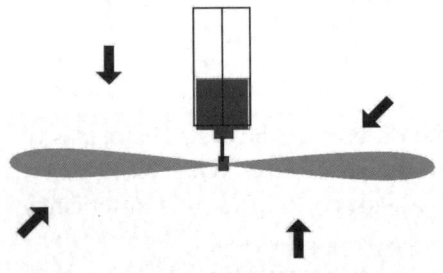

Figure 6.21: Blowing at the wind rotor from differenct directions

6.15 What to do when there is no natural wind

Note
For the experiments in the tutorial kit the wind should be the primary energy source.

How to produce wind

A simple and practical method to produce wind is the use of an electric fan. However, the air flow caused by such a device is much weaker than natural wind and you only get weak results. Nonetheless, you can experiment with such a setup. When you do so, it is interesting to try out different distances between fan and wind rotor and different locations of the fan (in front of or behind the rotor or sideways). Most fans allow you to control the rotation speed. However, the air stream produced by most devices of this type is not sufficient for meaningful experiments.

Most of the wind power experiments in this tutorial kit can be performed by blowing at the rotor blades.

It is also possible to take the bottle with the wind rotor in your hands and to pirouette or to run around. The slip stream is sufficient to turn the rotor.

After having experimented with these simulations it can be interesting to wait for natural wind.

Figure 6.22: Use a fan to produce wind

Figure 6.23: Driving the wind rotor by moving the bottle from the right to the left and utilising the slip stream

6.16 Wind speed and energy

Required parts: generator with wind rotor as before, breadboard, one red and one black alligator-clip cable each, moving-coil gauge

> **Note**
> For the following experiments you need sufficient natural wind or 'synthetic' wind.

Attach the wind rotor (consisting of generator, generator housing and blade assembly) to the bottle (half-filled with water for a firm standing). In order to use natural wind you should set up the whole construction outdoors in a place where you can feel a consistent wind. Alternatively, you can simulate natural wind and drive the rotor with a fan. In this experiment the switch in the circuit diagram represents a wire jumper to bridge the diode.

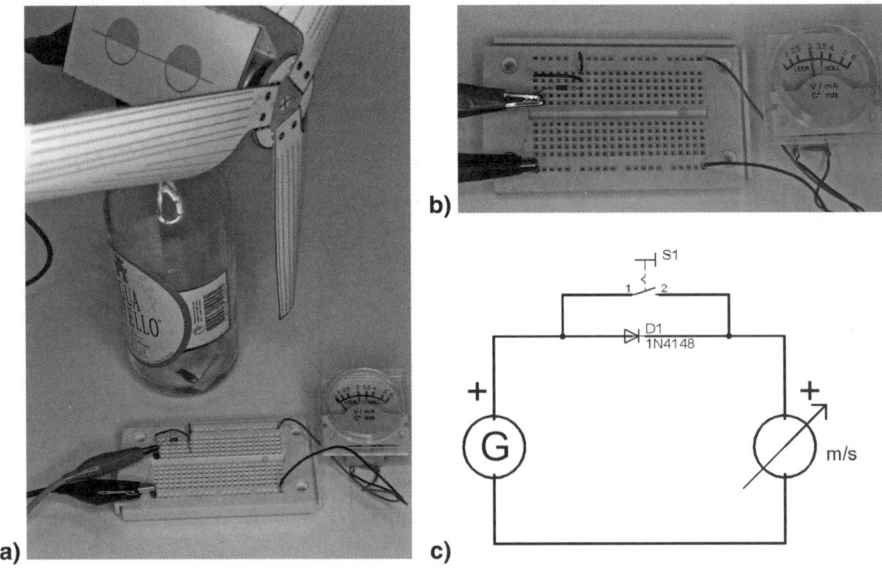

Figure 6.24: Anemometer (a) Setup (b) detail (c) circuit diagram

When you measure natural wind with the setup described above (open wire jumper) the needle deflection shows the wind speed on the scale in metres per second (m/s). Of course, the values can vary because the setup is not calibrated. If you have a calibrated anemometer at your disposal, you can use it to calibrate your own instrument.

With the setup described above, the wind speed is measured by aligning the wind rotor with the main wind direction. This is not necessary with professional anemometers because they have a vertical axis. The wind direction can be determined independently by using a vane and a direction sensor.

 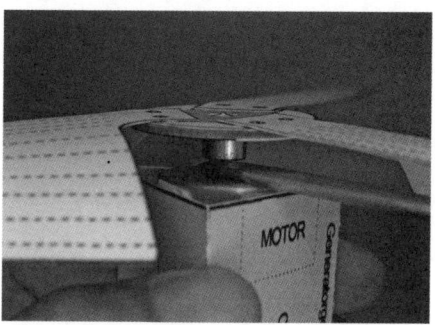

Figure 6.25: Anemometer consisting of plastic easter eggs cut in halves and a motor with bell anchor

Figure 6.26: Removing the wind rotor from the generator shaft

Removing the wind rotor from the generator shaft

Since the brass adapter ring is custom-fit, it is often sufficient to push it lightly on the generator/motor shaft (without tightening the grub screw). When removing, you should lever the adapter ring off the axis using a screwdriver. Fit the blade of the screwdriver between the brass part and the generator/motor casing and move it gently until the ring glides off the shaft.

7 Water Energy

The history of utilising water power goes back a long way. Even thousands of years ago water-driven scoop wheels were used to irrigate fields. In the early middle ages, the first water wheels in Europe were used for agricultural and mechanical purposes. To this end the wheels were equipped with drive belts and gear transmission. This made it possible to power mills, saw mills and forges. Today, water power has a huge share in the production of electricity by renewable energy.

7.1   What is water power/water energy?

The first question about water power is: What causes the usable power of water or water circuits? The basic natural motor is the sun that drives the natural water circulation.

The basic principle is as follows: The surface water on Earth is evaporated by the warm sunlight and gathered in clouds. The precipitation from the clouds, e. g. in the form of rain, drops down to the surface. It accumulates in higher regions and is guided by brooks and rivers into the valleys. The precipitation drained off by rivers can now be used to gain energy. In general, the kinetic energy of the water is used by collecting the water in large reservoirs and then converting the kinetic energy to electricity with turbines.

This explanation is of course a simplification. The complete water cycle is much more complex. Many other natural factors come into play, e. g. a sustainable vegetation, sound soil, naturally wooded hills, a minimum of sealed areas etc. The topic of 'renewable energy' thus implicates an examination of environmental issues.

A first and simple experiment shows an important part of the natural water circulation.

Required parts: a large bowl, a tissue handkerchief, wrapping film, a pebble, water.

Pour a little water into the bowl and put the tissue handkerchief inside so that it soaks up the liquid. Cover the bowl tightly with wrapping film. Place the pebble in the middle of the film so that it bends down slightly. Now place the whole setup in the sun. By warming up the water contained in the handkerchief evaporates and gathers underneath the wrapping film. There it condenses und drips down into the middle of the bowl, back onto the tissue handkerchief. This process is repeated over and over as long as enough solar radiation hits the bowl/ the wrapping film.

In Germany, water energy has a huge share in the production of energy out of renewable sources. In most of the modern hydro-electric power plants water flows through one or more turbines that drive a generator converting the mechanical energy to electricity. The contribution of this energy source to the production of electricity amounts to approx. 4 % in Germany and to more than 20 % all over the world and is yet extendable. Hydro-electric power plants may utilise the natural flow of water in a river. The water is collected in a reservoir and then guided through the turbines.

One of the benefits of water power is the fact that it can be stored easily. By damming up a flowing river and pumping the water onto a higher level (pumped storage power station) the energy can be stored easily and is yet readily available at short notice. Water power is therefore ideally suited to respond to the peak load demands of the power grid. The power of electricity (and thus the power of electricity gained by utilising water energy) is measured in watt. Usually the values are given in kilowatt (kW).

> **Rule of thumb for estimating the power output of a hydro-electric power plant**
>
> Height of drop in metres (m) x water throughput in cubic metres per second (m³/s) x 7 = electric power in kilowatt (kW)
>
> **Example:**
> Height of drop: 10 m
> Throughput: 500 l/s = 0.5 m³/s
> Power output: 10 x 0.5 x 7 = 35 kW

7.2 Utilising water power

Basically all water wheels can be used as a mechanical direct drive. Scoop wheels were already used in various cultures of the past, e. g. in Egypt, India and China. Historians have discovered that scoop wheels were in use in Mesopotamia as long back as 1200 B. C.

The Romans used water wheels to grind grain. From the 12th century onwards, water mills became prevalent in Europe. A little later, water power was also used for oil mills, saw mills and hammer mills (forges). During the beginning of industrialisation, machines were driven by water power. The power of large water wheels was transferred to the machines by means of transmission belts.

When water power is converted into electricity, it is no longer necessary to transmit the energy mechanically on-site. The electric current can be utilised independently of the location of the power plant. The conversion of water power to electric energy is completely emission-free.

7.3 Experiments on water power

The following practical experiments let you experience the properties and the operation of a water wheel. You will learn how water can be utilised and what you have to bear in mind in order to gain optimum energy yields. In the following text the electrical machine is called a *generator* because it is used to convert water power to electrical energy.

Adapting the generator and the generator housing

Required parts: generator, generator housing, yoghurt pot, double-sided adhesive tape

To make sure you can handle the generator safely and easily in experiments with water, you have to integrate it into a yoghurt pot. This way you reduce the chances of the generator getting wet.

The building instructions for the generator housing can be found in section 6.5, 'Building the generator housing'.

Cut a central hole (approx. 8 mm in diameter) into the base of the yoghurt pot using a small knife. Make sure the plastic material does not crack! Now attach some pieces of double-sided adhesive tape to the flat side of the metal generator casing. After removing the protective strips, insert the generator (together with its cardboard housing) into the yoghurt pot and stick it to the base. The drive shaft has to be centred in the base of the pot. Beforehand you have to clean the base of the pot thoroughly so that the adhesive tape sticks well to the plastic material.

> **Note**
> For the assembly of the generator housing see section 6.5, 'Building the generator housing'.

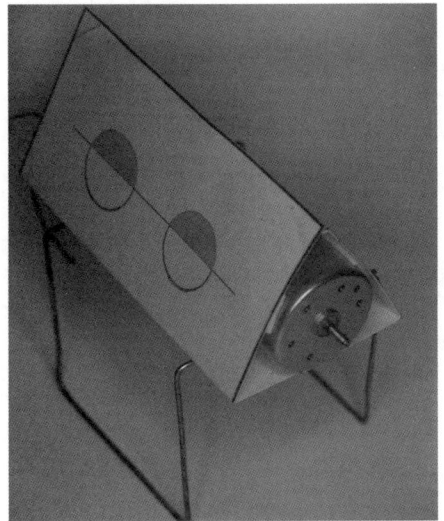

Figure 7.1: The generator
in its cardboard housing

a)

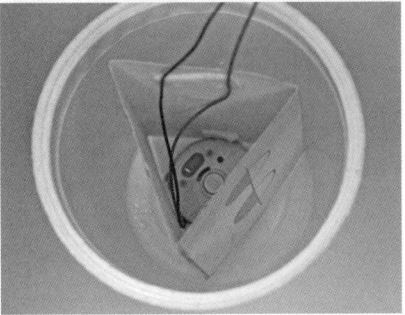

b)

c)

Figure 7.2: Mounting the generator
inside the yoghurt pot: (a) adhesive
tape on the metal casing (b) front
view (b) back view

The electrical machine has two connectors with a red and a black wire, respectively. The generator provides direct current and therefore has a plus and a minus pole, just like a battery. Connect the black and the red wire to the black and the red alligator-clip cable, respectively. Make sure the thin wires have contact with the alligator clips. It is best to bend the tin-plated wire ends so that the alligator clips have a good grip.

After attaching the wire ends to the alligator clips you can perform a function test with a red LED. Connect the LED to the free alligator clips and turn the generator shaft with your thumb and index finger. In one direction of rotation the LED should light up briefly.

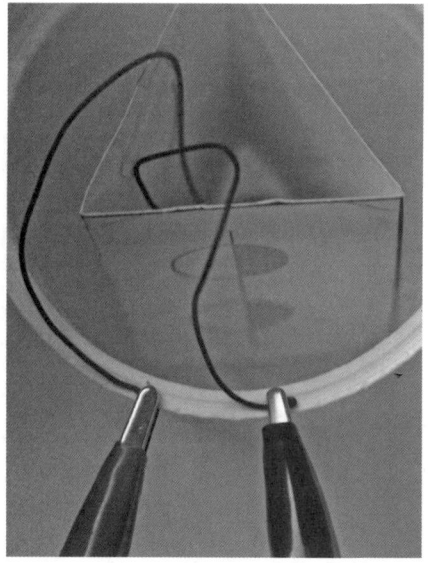 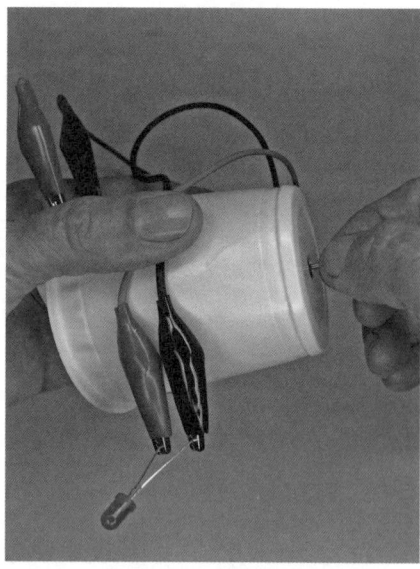

Figure 7.3: Connecting the generator wires to the alligator clips

Figure 7.4: A first test with an LED attached to the alligator clips

7.4   Types of water wheels

During the history of mankind, various types of water wheels have been developed. Basically you distinguish between *undershot* and *overshot* wheels. This denotes the position of the wheel in relation to the water flow. With an undershot wheel, the water passes underneath the wheel, with an overshot wheel, it passes above the wheel.

The stream wheel works exclusively according to the action principle (utilising the flow energy). Other types of undershot wheels apply a combination of the action and the reaction principle (utilising the elevation energy). Overshot wheels work exclusively according to the reaction principle and are still used in small areas of application. They provide a significantly larger torque than high-speed turbines. This is an advantage for mechanical drives but a drawback when they are used to turn a generator.

For industrial purposes we use turbines. They draw a lot of criticism due to environmental reasons since they require substantial interference with the natural water circulation. The construction of turbines is oriented on the output power and the basic parameters (like height of drop and pressure). The most commonly used types of turbines are:

○ Francis turbine
○ Kaplan turbine
○ Impulse turbine (Pelton turbine)

Francis turbines are universally applicable and most commonly used. The degree of efficiency can amount up to 90 %. Furthermore, it provides the advantage that it can be utilised 'in reverse' as a pump. Thus you can build a pump storage plant with just one turbine that is used to store water energy as well as to produce electricity.

Type	Description
Stream wheel	Flat blades are immersed in streaming water. The wheel utilises only the flow energy of the water (action principle). Due to its very low degree of efficiency, this type is rarely used today.
Undershot water wheel	With this type, water flows through an area around the lower third of the wheel and into the blades that are shaped in a hydrodynamically more efficient way. Therefore, the elevation energy (weight) of the water can also be used to a small account in addition to the flow energy.
Breast shot water wheel	This type is only used in isolated applications with small heights of drop. In the advanced shape of the Zuppinger wheel, it achieves a degree of efficiency of up to 75 % and is thus an alternative to turbines in small applications.
Overshot water wheel	Water is guided onto the wheel from above. This way, only the elevation energy, i. e. the weight of water, is utilised (reaction principle). Depending on height of drop, this type of wheel can achieve a degree of efficiency of up to 75 %. Flotsam screening as required for undershot wheels and turbines is not necessary because flotsam can simply run off the wheel.

7.5 Building the water wheel

Required parts: cardboard disc (flap wheel) and six flaps on the cut-out template, brass adapter ring

The detailed building instruction for the flap wheel that is used here as a water wheel is given in section 6.7, 'Building the drag-type rotor'. Additionally, you should impregnate the cardboard flap wheel with wax spray, hair spray or plastic spray so that it is more durable when experimenting with water.

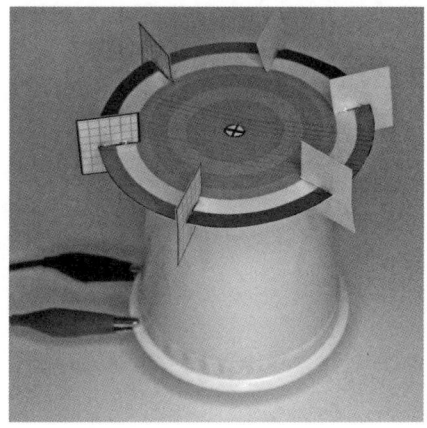

Figure 7.5: The flap wheel used as water wheel, mounted on the drive shaft of the generator

7.6 Experiments with the water wheel

Required parts: generator with attached water wheel

Install the water wheel on the generator shaft and direct a spray of water out of the tap onto the flaps. Try out several directions. In order to protect the generator it should be inserted in a yoghurt pot or the like. At first, perform the experiments without any electrical connection of the generator.

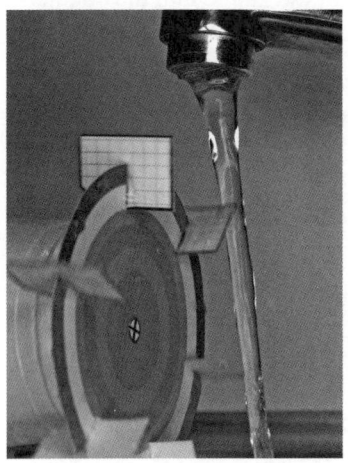

Figure 7.6: Trying out the water wheel

In the next step, fill the syringes contained in the tutorial kit with water, point the outlet towards the wheel and pop out the content.

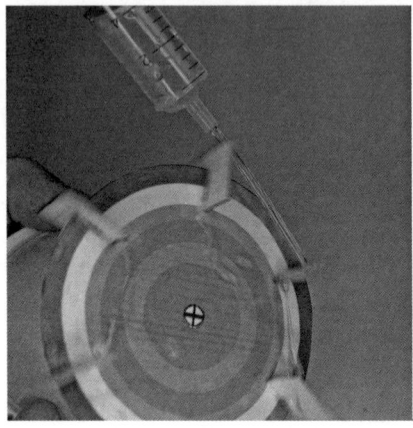

Figure 7.7: Driving the water wheel by water power out of the syringe

7.7   Preparations for measuring water power

Required parts: Breadboard, moving-coil gauge

The tutorial kit contains a moving-coil gauge that you will use in several of the following experiments. Insert the connection leads of the gauge into the contacts of the breadboard – the black wire (minus pole) into the lower bar and the red wire (plus pole) into one of the upper contacts.

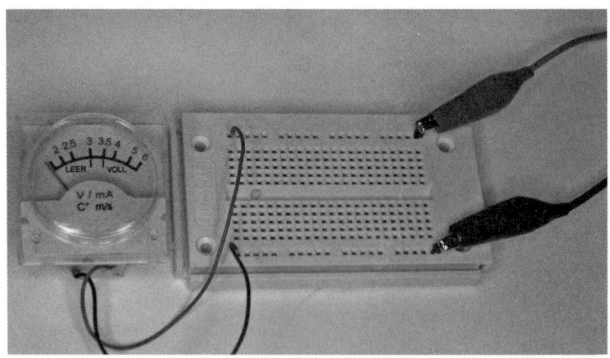

Figure 7.8: Connecting the gauge and the alligator clips to the breadboard

> **Note**
>
> Never connect the moving-coil gauge directly to a battery, because this could damage it.

7.8 Measuring the voltage of water-generated electricity

Required parts: generator with housing and flap wheel, breadboard, alligator-clip cables (one red, one black), two pins, moving-coil gauge

> **Note**
>
> Each experiment in this section builds upon its predecessor. Thus, it is not necessary to dismount the prior setup. Instead, you add, remove or replace individual parts to extend it.

Insert one pin into the lower and one into the upper bar of the breadboard. Attach the generator via the alligator-clip cables to the pins, for the time being without considering the polarity (minus/plus). Place the flap wheel under running water. Let the jet hit the flaps first on one side of the wheel and then on the other side so that the direction of rotation changes. Monitor the gauge.

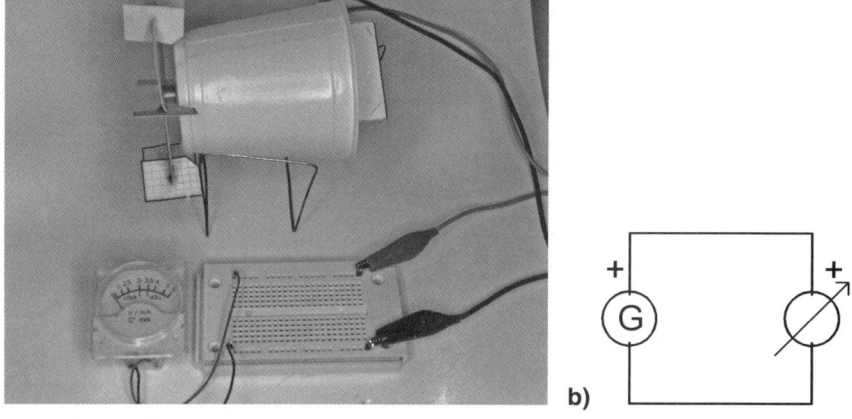

a)   b)

Figure 7.9: (a) Setup with flap wheel and gauge (b) circuit diagram

Depending on the polarity of the generator connection to the gauge and the direction of rotation, the needle of the gauge is deflected to the left or the right. When it is deflected to the left, the needle is blocked mechanically by the stop. In this direction you cannot read a value because of the missing scale. Therefore, you now extend the circuit by a diode so that the moving-coil gauge only receives electric current in the one direction that causes a usable needle deflection.

Required parts: as before, plus an additional silicon diode inserted according to the picture

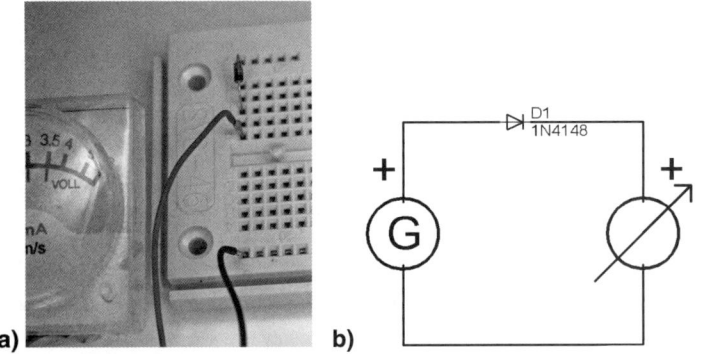

Figure 7.10: (a) Setup with additional silicon diode (b) circuit diagram

Now the needle of the moving-coil gauge is always deflected to the same side irrespective of the direction of rotation of the flap wheel. The direction of deflection depends on the polarity of the generator connection and the diode.

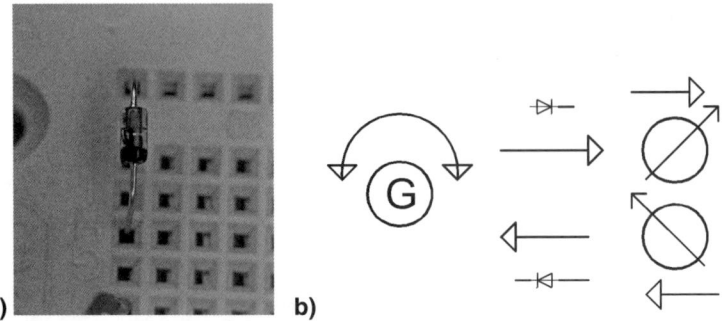

Figure 7.11: Mounting direction of the diode and its influence on the needle deflection

7.9   Measuring water power

Required parts: as before; generator with mounted flap wheel, connected to the moving-coil gauge via the black and red alligator-clip cables and the breadboard

You can perform this experiment with or without the diode. In the latter case, you have to direct the jet of water in such a way towards the wheel that the needle is deflected to the right.

It is best to perform the following experiment with a partner. One person can position the water wheel while the second one handles the tap and observes the gauge while the tap is opened more and more.

The more water flows onto the wheel, the more the needle is deflected.

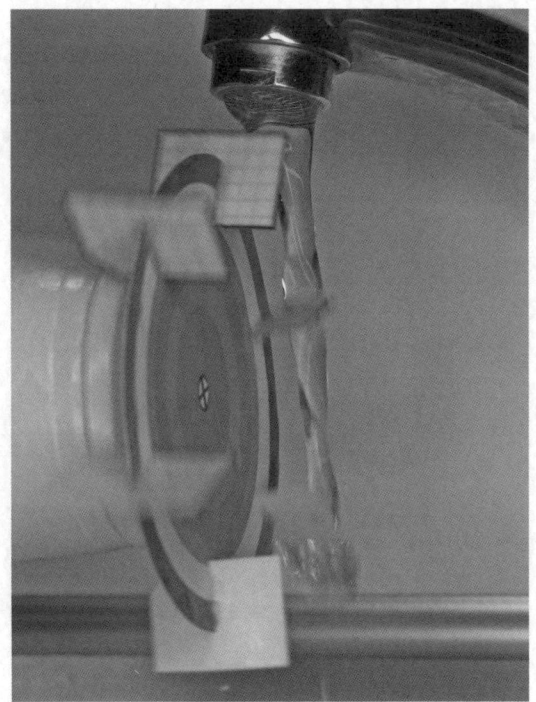

Figure 7.12: Measuring water power under an open tap

8 Storing Renewable Energy

We can harvest renewable energy when it is provided by nature. However, this happens irregularly and depends on the type of energy. For example, direct sunlight can only be utilised in the daytime and mostly in summer. Wind energy, too, depends on many facts. It can be utilised primarily in the night and in winter. The amount of usable water energy depends on the topography and the type of landscape.

It would be optimal to use energy directly, i. e. at the time when it is available. However, we also need energy at other times.

Therefore, a means of storing energy is required. Depending on the type of storage, the energy can be buffered more or less without losses for shorter or longer time periods and then withdrawn from the storage and utilised. There is a plethora of storage technologies. Some of the most important ones are discussed in the following sections. Afterwards, you can carry out some experiments on some of these storage types.

Summary of storage facilities

❶ Mechanical storage

- ○ Pump storage power plants
- ○ Compressed-air storage
- ○ Flywheel storage
- ○ Gas tank

❷ Thermal storage

- ○ Storage with liquid or solid medium
- ○ Latent heat accumulator

❸ Chemical storage
- ○ Rechargeable accumulators
- ○ High-temperature accumulators

❹ Thermochemical storage

❺ Electrochemical and magnetic storage
- ○ Electrolytic capacitors
- ○ gold caps

8.1 Experiments on storage technologies

The utilisation of renewable energy involves the problem that the amount of usable energy varies because it is influenced by natural systems. This is especially true for wind and solar energy.

In order to ensure that the consumers receive a regular energy supply by the power grid, the providers still have to maintain gas or coal-fired power plants. They can be turned up and down flexibly to cover the variations in the availability of renewable energy. To make better use of the varying yield of renewable energy it would be advantageous to use reasonable storage technologies. In the following sections you can read about some of the many possibilities for storing energy and study them by performing experiments.

8.2 Compressed-air storage

In the practical deployment of wind power, we use pump storage systems. Besides that, there is some research going on to put the compressed-air storage of wind-generated electricity into practise.

As opposed to water-driven pump storage plants, the excess energy in this type of storage is used to compress air in a storage tank. When there is a high demand for energy, the compressed air is used to drive a special turbine. Salt caverns can be used as storage tank for huge amounts of energy. The drawback is a comparatively low degree of efficiency, i. e. an unfavourable ratio between the stored and the withdrawn energy.

Required parts: syringe contained in the tutorial kit

You can experience the principle of compressed-air storage with the help of a syringe. Pull out the plunger to the 10 ml mark on the scale. Seal the outlet of the syringe with a finger of your other hand. Now push in the plunger against the rising pressure as far as possible. Release the plunger, and it slides out by itself. This is the most basic form of compressed-air storage. The mechanical energy of your hand is stored and released again when you let go of the plunger. You can clearly see that the plunger does not slide out far enough to reach its original position. This shows that the withdrawal of energy has a degree of efficiency less than 100 %. After releasing the plunger, you can read a value between 85 and 90 on the scale.

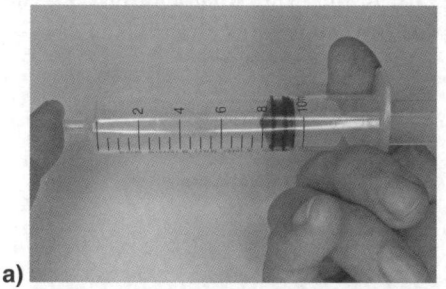
a)

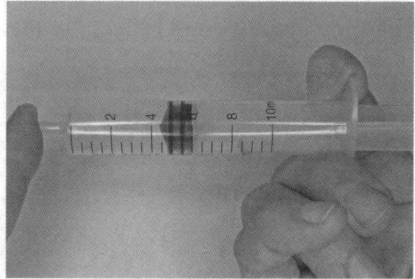
b)

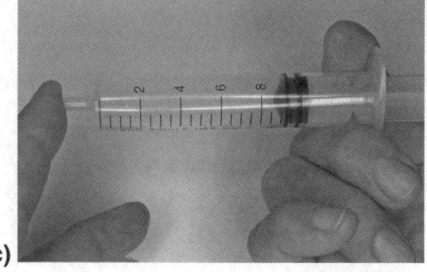
c)

Figure 8.1: A syringe as a basic compressed-air storage: (a) handling (b) compressing air (c) scale after releasing the stored energy

8.3 Chemical storage

Accumulators like those that are used as an alternative to single-use batteries represent a typical form of chemical storage. There are various types of accumulators that differ mainly in the chemical components and the inner structure.

Energy providers and battery manufacturers try to find solutions to store wind and solar energy in huge accumulators. Megabatteries are meant to buffer electrical power up to the megawatt range (MW) so that it can be fed into the power grid at a later time.

Charging an accumulator

Accumulators are common energy storage devices to drive electrical and electronic devices. They are also used as a replacement for expensive single-use batteries. The technical process used to charge the accumulator influences the charging capacity and especially the service life (number of charging cycles).

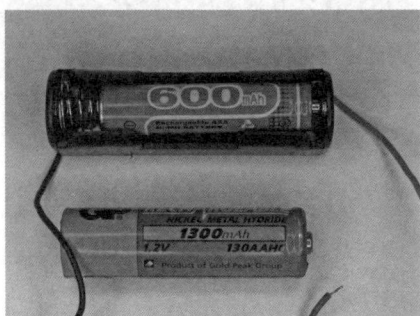

Figure 8.2: Accumulator types like microcells (AAA) and mignon cells (AA) are sutiable for many portable electronic devices

Constant current charge

Required parts: solar module, breadboard, resistor, LED, AAA accumulator, accumulator bracket made of bent wire

> **Note**
> For the following experiments you need a bright light source (or full direct sunlight) for the solar module.

The most basic charging technology is the constant current charge. The accumulator is charged with a given electric current for a given time period. With simple constant current charges, you normally charge the accumulator for 14 hours with 1/10 of the current specified in the capacity rating.

> **Example**
> Accumulator capacity 500 mAh, charging current 50 mA, charging time 14 hours. There may be an electronic circuit that switches to conservation charging after 14 hours. For this, 1/20 of the specified capacity can be used, i. e. 25 mA in this example.

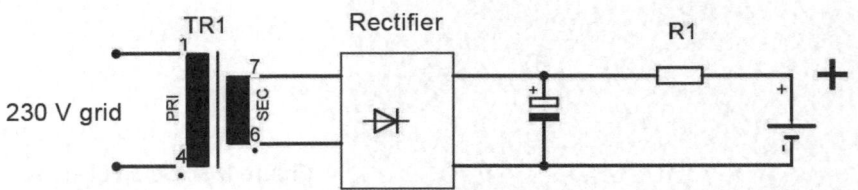

Figure 8.3: Conceptual circuit diagram of a constant current charger. The charging current is adjusted by choosing the right size of resistor.

However, using the resistance value to adjust the charging current would not be wise with a solar charger. Here you perform the adjustment in a lossless way by choosing the right size for the solar cells or the module.

With the right size of the module you do not even need a series resistor. When a solar module provides an electric current of 30 mA in brightest sunlight (as the module contained in this kit does), that accumulator in the preceding example can not be damaged in the charging process.

This changes of course with bigger (and more powerful) solar modules that provide a higher current. With these devices, a charge current limiting or a control electronic is mandatory, otherwise the accumulator will be damaged.

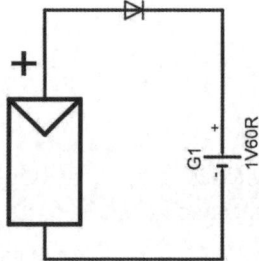

Figure 8.4: Circuit diagram of a simple solar charger

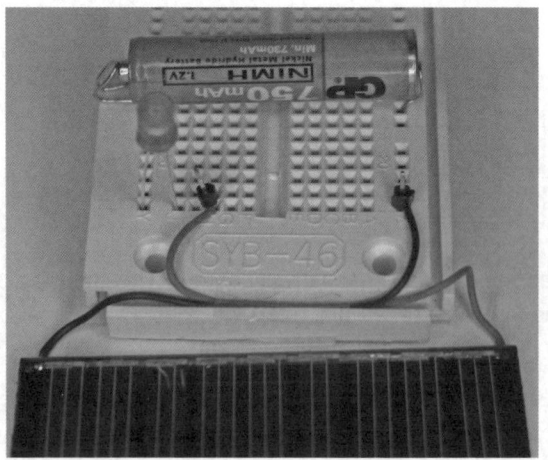

Figure 8.5: Setup of an experimental solar charger: the presence of a charge current is indicated by an LED that also functions as a non-return valve so that the accumulator is not discharged at night via the solar module.

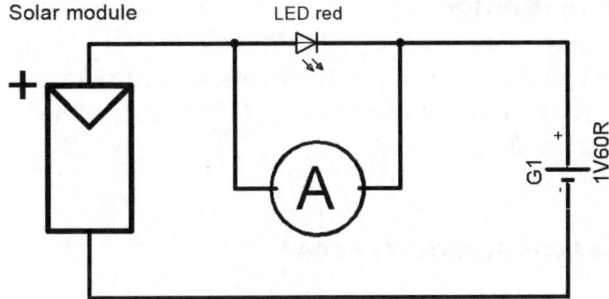

Figure 8.6: Measuring the charge current with the moving-coil gauge. With this circuit, the LED does not light up as long as the measuring device is connected.

Depending on the type of accumulator, there are several ways to adjust a solar module so that the accumulator is not damaged in the charging process. When using small NiCd or NiMH (nickel metal hydride) accumulators, the simplest solution is to control the maximum charge current provided by the solar module.

In contrast, lead acid and lead gel accumulators are controlled in the simplest way by the value of the end-of-charge voltage.

A 'large' solar lead accumulator with a voltage of 12 V can therefore be charged by a solar module with a maximum cell voltage (off-load voltage) of 15 V without any problems. The charging process controls itself. The more the charge voltage of the given accumulator increases, the lower the current provided by the solar modules becomes (automatic adjustment). Although this is a viable charging process, it is not optimal with respect to the usability the service life.

Lithium accumulators usually prove themselves unproblematic as long as the charge current and the accumulator capacity stay inside the limits and the maximum voltage of the solar module (approx. 4.2 V for one cell) is not significantly exceeded. Generally, these accumulators have a built-in electronic control circuit to protect them against low voltage and excess voltage.

8.4   Electrochemical storage

Electrolytic capacitors and gold caps belong to the electrochemical storage devices and provide an energy efficiency of more than 90 %. They can be charged
quickly, but offer a low energy density.

Electrolytic capacitors as storage devices

*Required parts: solar module, breadboard, 100-Ω series resistor (brown-black-
brown), red blinking LED, 4700-μF electrolytic capacitor*

> **Note**
>
> These experiments can be carried out in low light conditions as well (shad-
> ow, clouded sky).

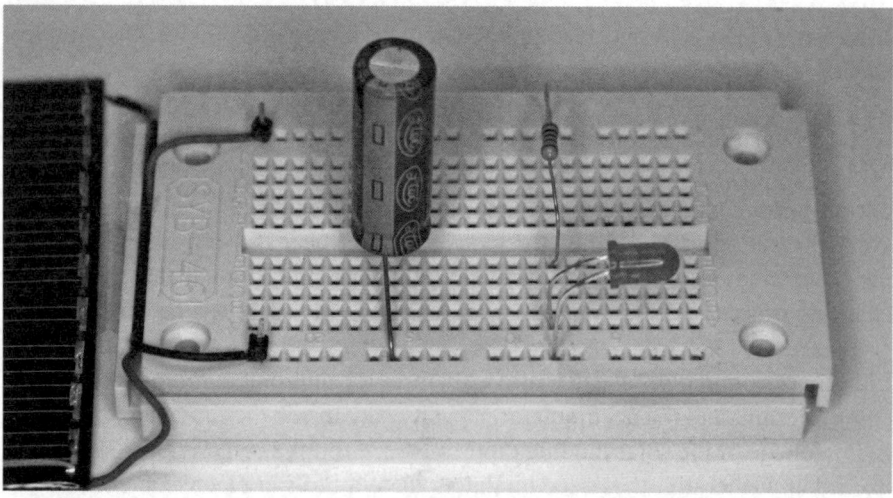

Figure 8.7: Setup on the breadboard with flashing LED. Insert the electrolytic capacitor while the LED flashes. What happens after you do that? The LED ceases to
glow. It takes some time until the LED lights up or flashes again. When you cover
the solar module, it flashes.

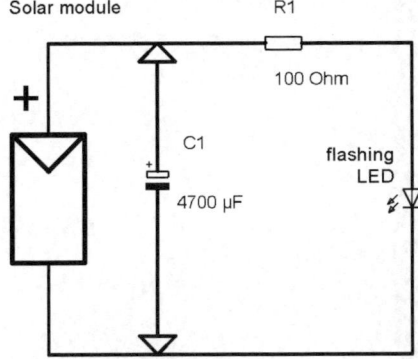

Figure 8.8: Important: Remember the series resistor R1 when connecting the LED!

Additional effect

In low light, the LED glows brighter with a capacitor than without it, although light source and solar module are not changed. The reason for this is that energy can be stored in the capacitor when the flashing pauses. This energy can then be released to the LED when flashing.

Performing the experiment: Use the setup in Fig. 8.7. Leave the electrolytic capacitor in the breadboard until the LED flashes. Remote the capacitor from the breadboard. Next, shade the solar module. The LED immediately stops flashing. Now insert the capacitor again into the same contacts as before while the solar module is still shaded. The LED flashes although no current flows in from the solar module.

Result: The charge in the storage capacitor is maintained for some time.

When the capacitor is loaded, the LED flashes. Remove the solar module. Determine how long the LED flashes while it is supplied only by the capacitor. The larger the storage capacitor, the longer the LED flashes without power supply from the solar module. With a gold cap (as used in the following section) the missing power supply from the solar module can be bridged for a longer time (e. g. at night).

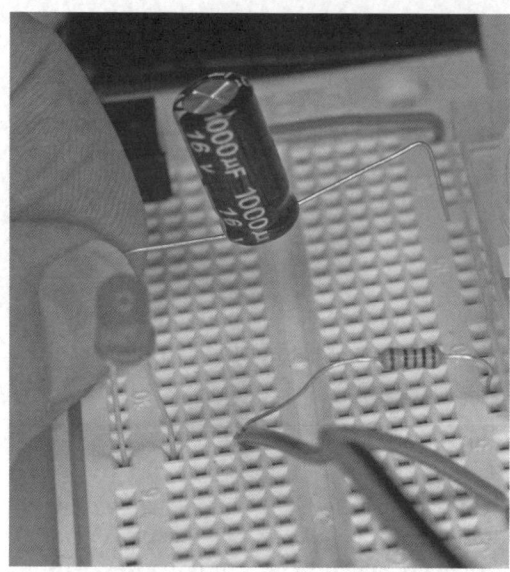

Figure 8.9: Remove the capacitor during the experiment and re-insert it later.

Storage in a gold cap

Required parts: solar module, breadboard, gold cap, 100-Ω series resistor (brown-black-brown), red flashing LED; additional orange LED for the extended circuit

---

Note

These experiments can be carried out in low light conditions as well (shadow, clouded sky). However, the charging process takes more time than in bright sunlight.

The experiment described in the last section can also be carried out with a gold cap instead of a normal electrolytic capacitor. When charging the gold cap, you have to make sure that you do not significantly exceed the admissible upper voltage limit. To avoid this, we will use a simple charge protection circuit further down.

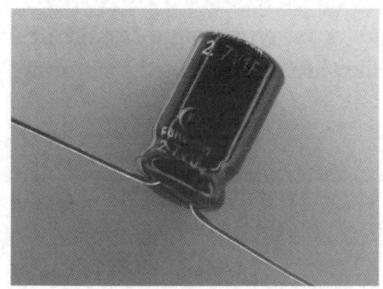

Figure 8.10: gold cap with a capacitance of 1 F and a maximum voltage of 2.7 V

Gold caps are electrolytic capacitors with a very high capacitance. Their service life is as long as the service life of normal electroiytic capacitors. It lasts some 100,000 cycles. In specialist electronics shops you can purchase gold caps with usual capacitances of 0.022 F to 22 F and operating voltages between 2.3 and 2.7 V. Gold caps with a capacitance of more than 2000 F are also available.

Gold caps are ideally suited as buffer elements when using solar energy for electronic applications – as long as they have a low resistance value. They do not need a charge control because they limit the charge current automatically by an internal resistor und thus cannot be overcharged. Furthermore, even deep discharging and short-circuiting cause no problems. The complete self-discharge time at room temperature amounts to 30–50 day.

However, the imprinted voltage should not be exceeded (or at best slightly). When a voltage of 2.3–2.7 V is specified, the capacitor has to be charged with no more than 2.8 V to be completely charged. The voltage limitation can be implemented in the most simple way, e. g. by a Zener diode connected in parallel. When charging with the given solar module, the end-of-charge voltage is adjusted automatically.

Gold caps can be connected in series and in parallel without any problems. A series connection increases the voltage, a parallel connection the capacitance of the whole setup.

Gold caps are usually deployed in solar-driven clocks, programme storage devices and solar-driven measuring instruments. They are ideally suited for storing solar energy.

To calculate the bridging time of a Gold cap in a buffering application, you use the following formula:

$$T = \frac{(V_1 - V_2) \times C}{I}$$

T = bridging time in seconds

V_1 = charge voltage in volts

V_2 = admissible maximum voltage of the device in volts

I = current consumption of the device in ampere

C = capacitance of the gold cap in Farad

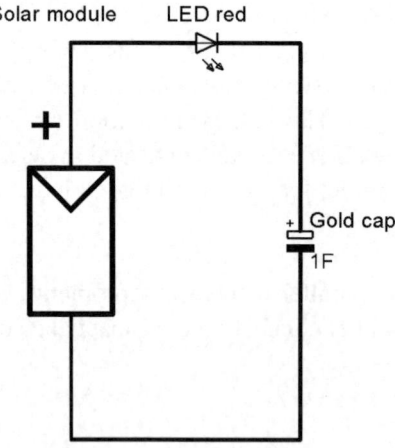

Figure 8.11: Gold cap storage. The LED serves as a charge indicator and prevents that the gold cap is discharged via the solar module in darkness.

Figure 8.12: Setup on the breadboard: a gold cap as storage device

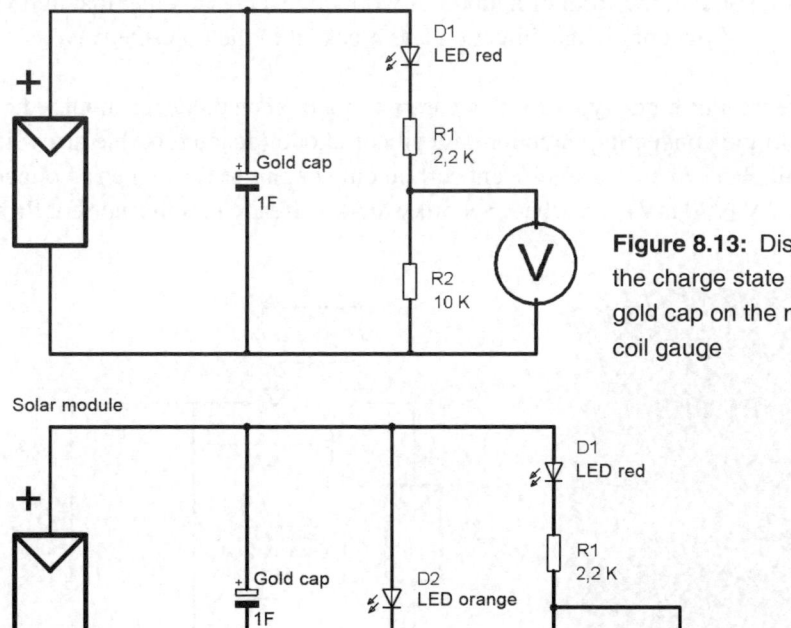

Figure 8.13: Displaying the charge state of the gold cap on the moving-coil gauge

Figure 8.14: Simple circuit with additional orange LED as overvoltage protection

The orange LED provides basic protection because when the voltage is increased, it consumes more current. Thus the current flowing from the solar module to the gold cap is reduced. However, this only works fine as long as the current provided by the solar module corresponds to the current consumption of the LED.

8.5 Discharge protection diode

Required parts: as before, but with an additional diode

The charge of an energy storage device like an electrolytic capacitor, a gold cap or an accumulator would dissipate via the generator. This can be prevented by a non-return valve in the form of a diode. It works like a check valve that allows the energy to flow only in one direction and blocks it in the opposite way.

There are various diode types for this purpose, e.g. silicon diodes, Schottky diodes etc. In the conducting direction of a silicon diode (denoted by the arrow in the circuit plan symbol), a significant current only begins to flow when a voltage of 0.6–0.7 V (700 mV) is reached. Schottky diodes already permit a current flow at 0.25 V.

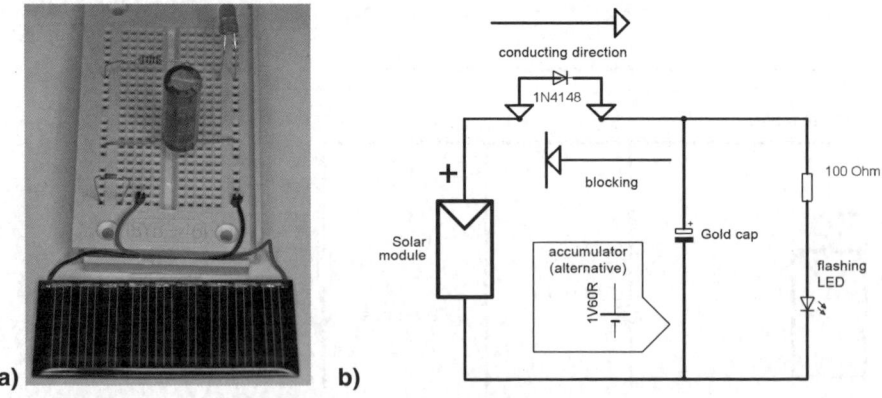

Figure 8.15: (a) Charging circuit on the breadboard (b) circuit diagram

> **Note**
>
> Blocking diodes prevent the discharging of the storage. Normal silicon diodes (blocking diodes) 'eliminate' approx. 0.6 V of the gained voltage, Schottky diodes only 0.25 V.

8.6 Storing wind and water power

Required parts: generator with wind rotor or water wheel, breadboard, two pins, 4700-µF electrolytic capacitor, push-button, moving-coil gauge

After assembling the components, let the generator – driven by wind or water – charge the storage for some minutes. Then you can test if the electrolytic capacitor has been charged by pressing the push-button. The moving-coil gauge should still display a voltage although the generator is no longer operating. The gauge acts now as a charge indicator und shows the charge state of the energy storage.

The displayed value decreases while you press the button because the small charge of the capacitor is further discharged via the gauge.

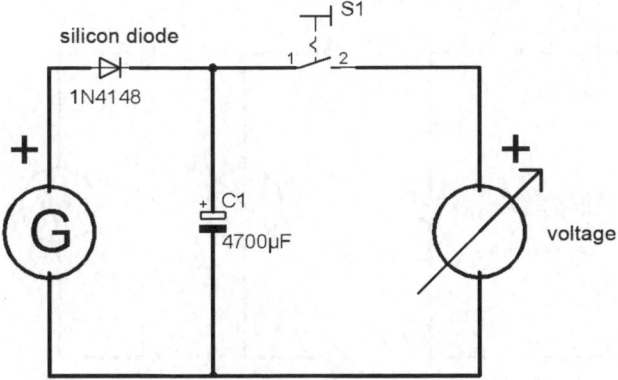

Figure 8.16: Measuring the charge with the moving-coil gauge

8.7 Demonstrating the charging operation

Required parts: generator/motor, a red and a black cable with alligator clips, 4700-μF electrolytic capacitor, 10-kΩ resistor (brown-black-red), red LED, battery clip, 9-V battery, moving-coil gauge

The charging circuit can be used to perform the preceding experiments without wind or water in order to experience the basic charging and discharging operation.

a)

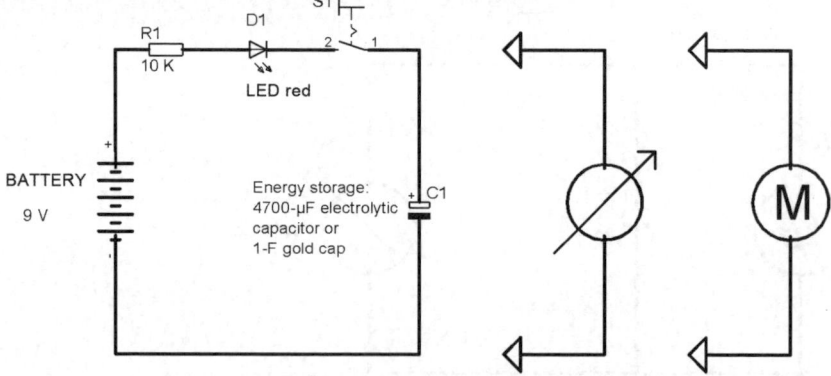

b)

Figure 8.17: (a) Setup of the charging circuit on the breadboard (b) circuit diagram

Observations

The electrolytic capacitor is charged with a current reduced by the resistor. The red LED indicates the charge current. It glows as long as the charge current flows, and the more the capacitor is charged, the brighter shines the LED. When the charging process is finished, the LED goes out.

When you connect the motor or the moving-coil gauge, the storage capacitor is discharged and thus the LED lights up again.

8.8 Solar storage system

Required parts: solar module, breadboard, red LED, gold cap, 10-Ω series resistor, flashing LED, silicon diode, four pins, push-button, motor, e. g. with flap wheel

Insert all components into the breadboard or connect them, respectively, according to the picture. Charge the gold cap via the solar module, using a light source like e. g. the sun. The charging process is indicated by the red LED. The flashing LED begins to flash when the gold cap is charged completely. When you press the button, the stored energy is released to the motor so that the drive shaft turns.

With this setup you can experience the whole chain of operation from charging the energy storage by renewable solar electricity to the consumption.

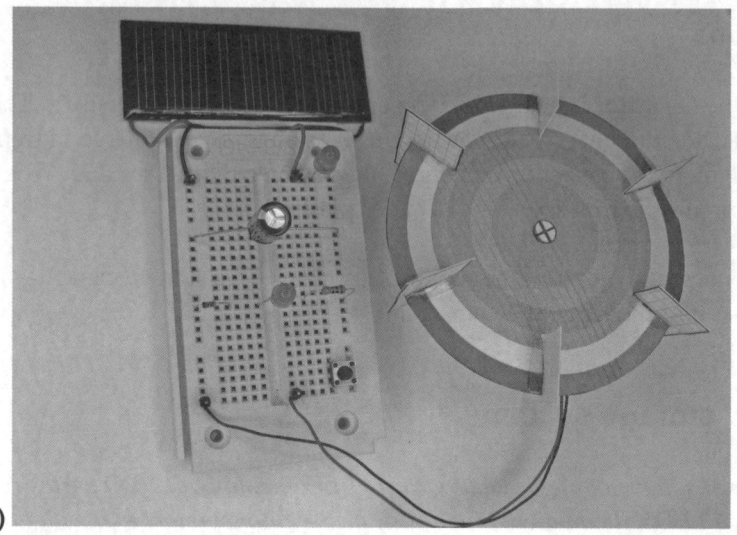

Figure 8.18: (a) Setup on the breadboard with solar module and motor (b) circuit diagram

8.9 Combining wind power, water energy and solar electricity

Required parts: generator with wind rotor, one red and one black cable with alligator clips, solar module, breadboard, gold cap, push-button, 1-kΩ resistor (brown-black-red), orange LED, 100-Ω resistor (brown-black-brown), diode, battery clip, moving-coil gauge

The following setup shows how to charge the storage device (gold cap) alternately with electricity gained by wind power, water power or other renewable sources of electrical energy like photovoltaics. The third source of energy in this experiment is the solar module.

Insert all components into the breadboard. Attach the generator via the red and black alligator-clip cables. This time, you connect the red clip to the lower bar of the breadboard (minus pole) and the black clip to the upper row of five contacts. Also connect the solar module to the breadboard. Now the orange LED should light up brightly. It will glow until the energy storage (gold cap) is completely charged. Afterwards, the LED goes out because no more charge current flows to the storages. Now you can press the push-button. The wind rotor begins to turn and the needle of the moving-coil gauge is deflected far to the right (showing the charge state of the gold cap). Then the deflection goes back slowly.

When you charge the gold cap with wind power, you will observe the following: When you blow at the wind rotor, the needle of the gauge is deflected. The electrical energy is guided via the diode to the gold cap and stored. The gold cap acts as an energy storage for wind and water power. In the next step you can use natural wind, a fan or water power to charge the energy store until the energy is displayed on the moving-coil gauge and the motor is driven.

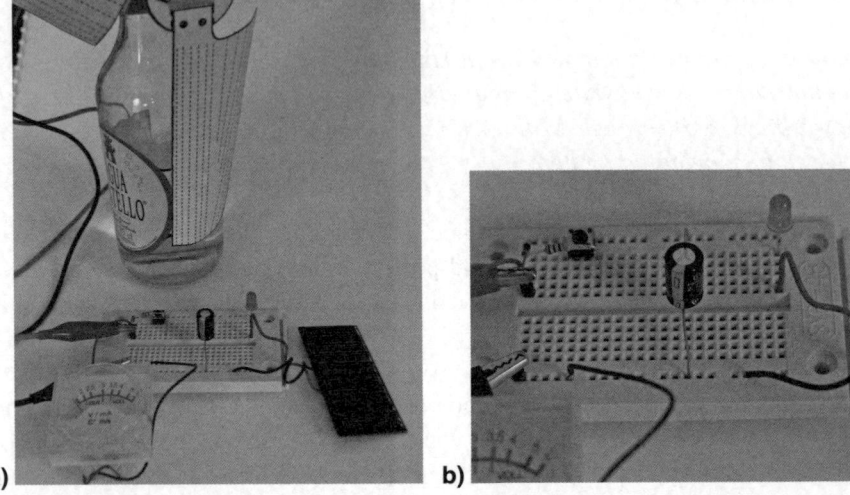

Figure 8.19: (a) Setup with wind rotor, breadboard and moving-coil gauge (b) setup on the breadboard

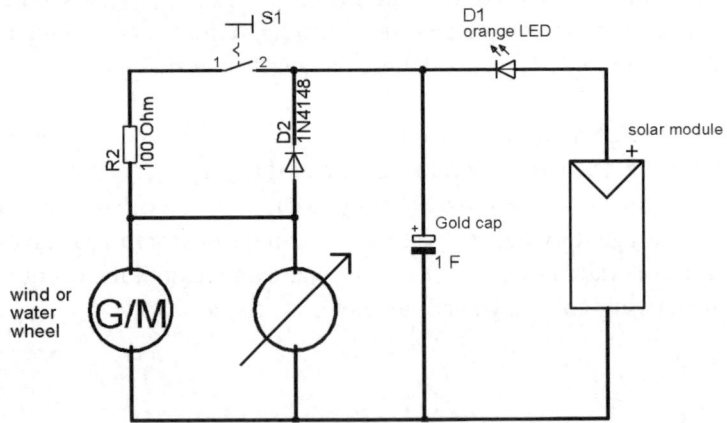

Figure 8.20: Circuit diagram, suited for using the wind rotor or the water wheel

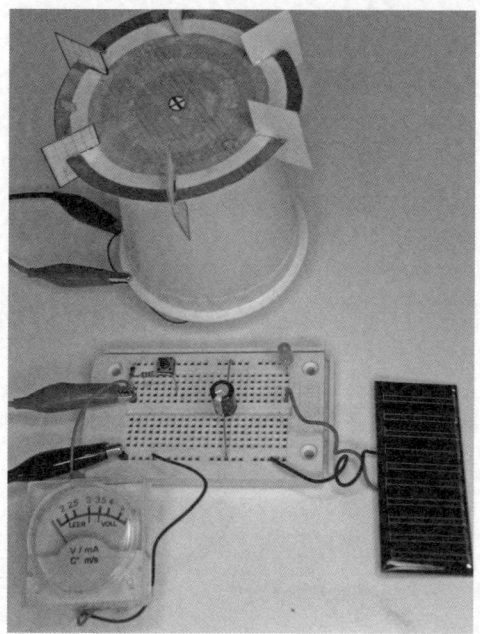

Figure 8.21: Setup with water wheel and solar module

Appendix

A.1   Test circuits and tests

Electronic components are generally very robust, but they can be destroyed by improper handling. On the following pages you will find test circuits for LEDs, transistors and ICs so that you can test the proper operation of these components.

A good contact of the wires in the breadboard is as important in the test circuit as it is in the experiments. It may happen that the connection lead of a component or a jumper wire is not inserted deeply enough into the contact bar. In this case, the circuit will of course not work properly.

The LEDs and the transistor can be tested with the solar module included in this kit. For testing the IC, you need a 9-V battery.

Testing LEDs

When an LED does not glow, the possible cause may be that the connection leads were inserted the wrong way round. If you connect an LED by accident without a serious resistor to a solar module in full sunlight, to the battery or a fully charged electrolytic capacitor, you see a short flash of light, and then the LED is destroyed. When this happens to one of the normal LEDs, it is not the end of the world since you have two of them (the red and the orange one) and you can replace the destroyed LED by the other. However, when you kill the flashing LED, you have to purchase a replacement to go on with the experiments.

For testing an LED you have to setup the circuit described below on the breadboard.

Required parts: solar module, breadboard, 1-kΩ series resistor (brown-black-red), the LED to be tested

Note

This test works best with a point light source like a desk lamp, but you can also resort to sunlight under a clear or lightly clouded sky (so you can see whether the LEDs glow).

Attach the plus pole of the solar module to the upper bar of the breadboard and the minus pole to the lower bar. Connect the plus pole bar via a 1-kΩ series resistor to one of the rows of five contacts and the minus pole via a piece of wire to another row. Insert all LEDs into the breadboard. When none of the LEDs contained in the tutorial kit works (the red one, the orange one and the blinking LED), it is likely that the circuit is wrong or the LEDs were inserted the wrong way round. When the circuit is working properly and one of the LEDs does not work although it is inserted correctly, the component is defect and should be replaced.

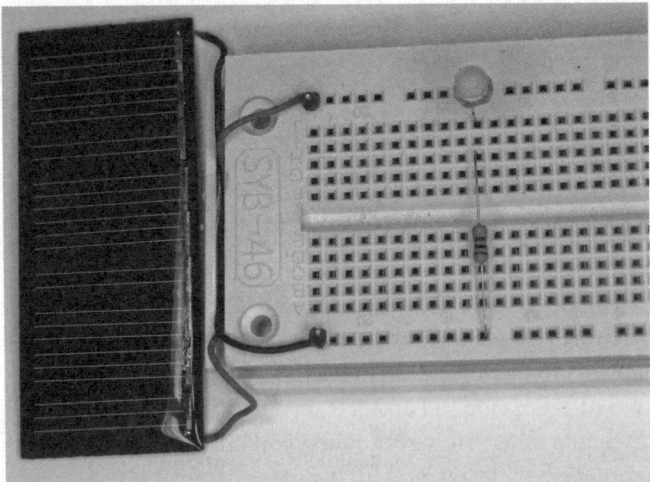

Figure A.1: Setup with solar module and LEDs. The longer connection lead of the LED is the plus pole.

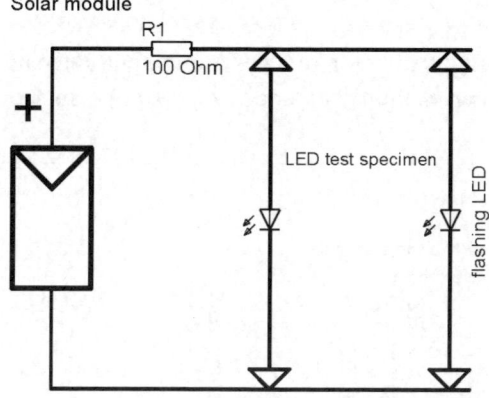

Solar module

Figure A.2: Test circuits for LEDs. In order to test the circuit you insert all available LEDs in the first step. Then you insert only the LED to be tested.

Testing the transistor

First make sure that the connection leads are attached correctly. The base connection is in the middle, emitter and collector are located at the sides. To distinguish between them correctly, you have to know whether you look at the front side of the transistor (flat) or the back side (bulged). When you look at the type specification, the emitter is the left leg and the collector the right one.

Most problems can be solved by determining the identity of the connection leads so that you can attach the transistor in the proper direction. If not, you can test whether the component works with the procedure given below. However, for this, you need working LEDs.

If the transistor is not working, you will have to replace it.

Testing an NPN transistor

Required parts: solar module, breadboard, 4700-µF electrolytic capacitor, 10-kΩ resistor (brown-black-orange), 1-kΩ resistor (brown-black-red), LED or blinking LED, push-button, the transistor to be tested

> **Note**
>
> This test works best with a point light source like a desk lamp, but you can also resort to sunlight under a clear or lightly clouded sky (so you can see whether the LEDs glow).

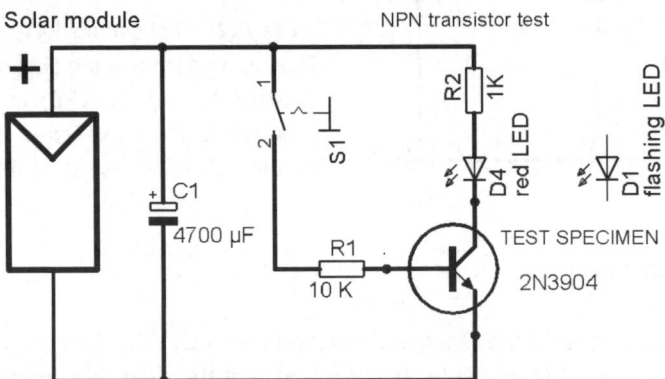

Figure A.3: Diagram of an NPN transistor test circuit. When you depress the push-button, the test specimen receives a base current. This should fully activate the collector-emitter section so that the LED lights up.

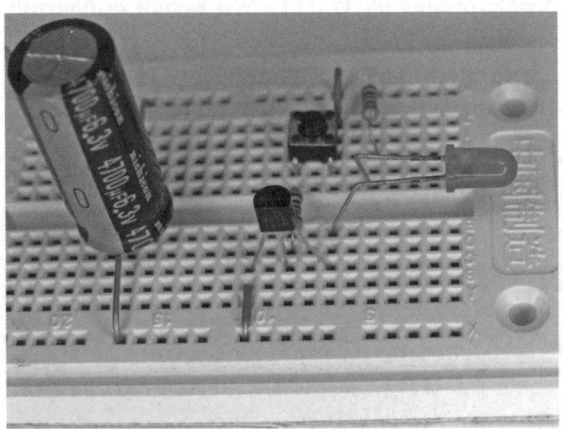

Figure A.4: Setup of the NPN test circuit on the breadboard. When you press the button and thus close the connection towards the plus pole, the LED should light up. If not, the transistor is defect.

When you press the button, the LED should light up. If it does not, the transistor may be defect.

Testing the LM 358 IC

The following simple circuit allows you to test the operation of the IC. The IC casing holds two distinct op-amps. You can test both independently.

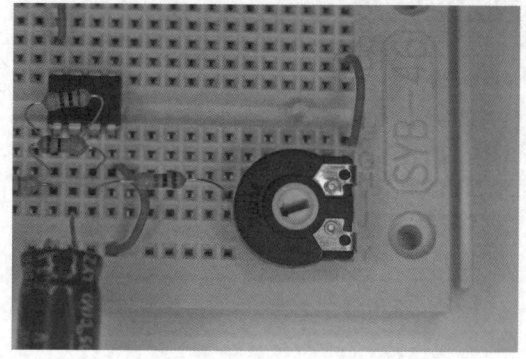

Figure A.5: Test circuit for the op-amps (see also Fig. A.7)

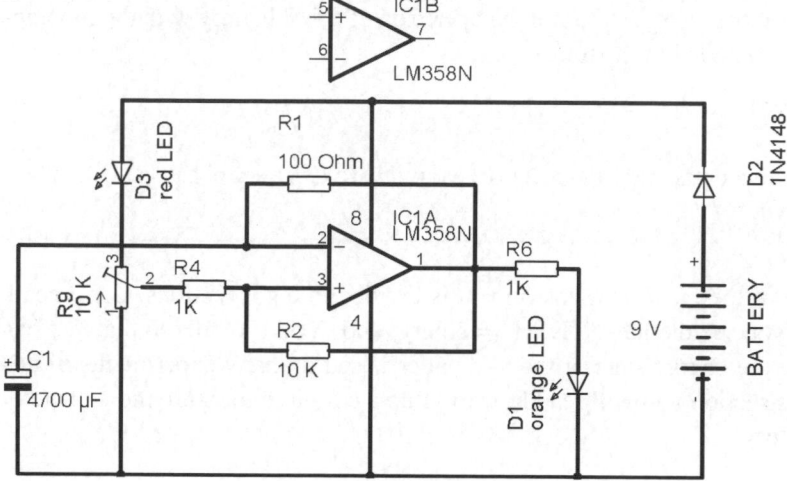

Figure A.6: Diagram of the test circuit

The test circuit shows the proper operation of the IC by the flashing of the orange LED. Depending on the setting of the potentiometer, the LED will flash only once, flash regularly or glow continuously. Thus the test circuit is also suited as an interesting blink circuit. The glow of the red LED (D3) indicates that the power supply is operating.

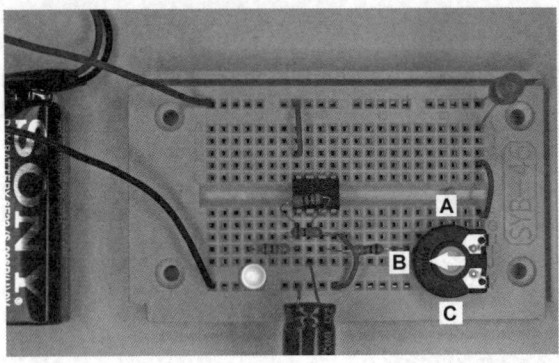

Figure A.7: Different settings of the potentiometer and their effects: (A) continuous light (B) steady flashing (C) short flashes

Testing the battery

You can test the charge state and the operation of a 9-V battery with the moving-coil gauge included in the tutorial kit.

> **Note**
> Never connect the moving-coil gauge directly to a battery because this could damage it.

The scale of the measuring instrument is divided in eight sections. Underneath the scale you see the labels 'Leer' (= empty) and 'Voll' (= full) so that you can determine the charge state of the 9-V battery used in many experiments of this kit. In this section you will read how to setup a circuit to measure the voltage of a 9-V battery.

Required parts: 10-kΩ resistor (brown-black-orange), moving-coil gauge, battery clip, battery, push-button, breadboard

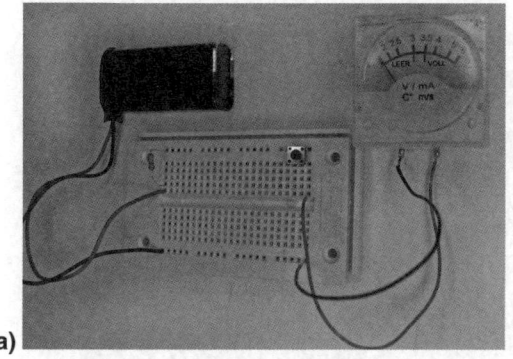
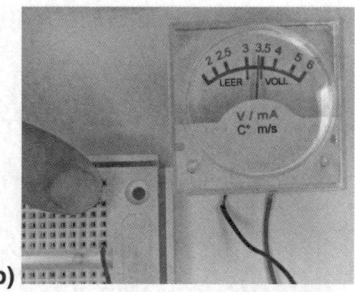

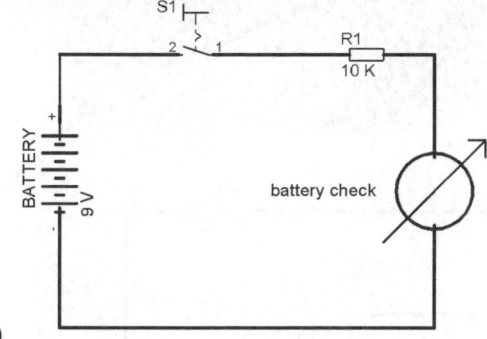

Figure A.8: (a) Setup
(b) needle deflection when
depressing the push-button
(c) circuit diagram

This circuit allows you to check the state of the battery. When the needle ends up in the 'leer'/empty section, the battery is no longer suited for the experiments. With a new, fully charged battery, the needle will be deflected up to the 'voll'/ full section. In the middle field between these sections the battery is still usable. Since the measuring instrument only needs a very low current for displaying the voltage, the test circuit shows the off-load voltage. For a more reliable statement on the state of the battery you have to measure the voltage under load. To this end you have to modify the circuit as follows.

Required parts: two 1-kΩ resistors (brown-black-red) connected in series (= 470 Ω), one 10-kΩ resistor (brown-black-orange), moving-coil gauge, battery clip, 9-V battery, push-button, motor or orange LED, breadboard

Modify the existing circuit according to the picture and the circuit diagram.

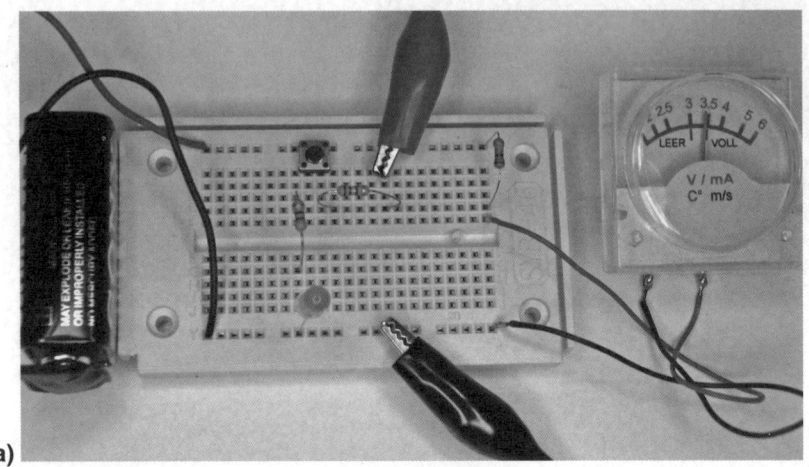

a)

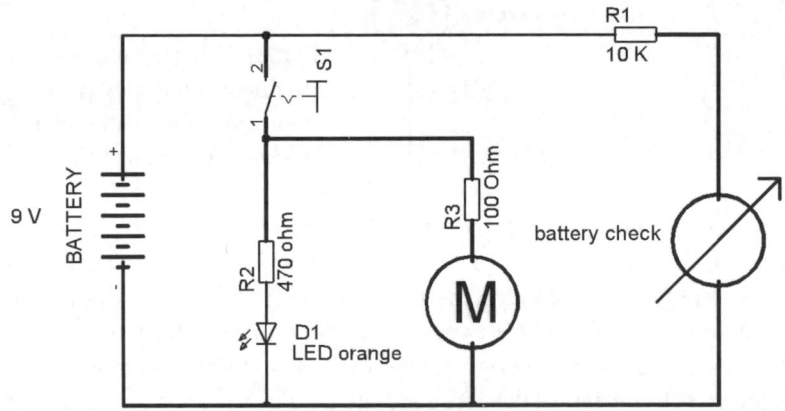

b)

Figure A.9: (a) Setup for testing the battery under load (b) circuit diagram

When you attach the battery to the battery clip, the off-load voltage of the battery is displayed immediately. By depressing the push-button you connect a consumer (LED or motor) in parallel to the battery. Thus you can measure the voltage with and without load. The deflection of the needle on the moving-coil gauge is reduced when you depress the button. The value displayed now is the voltage under load. This gives a more realistic picture of the charge state of the battery.